Max Parsonage

AS CHEMISTRY

FACTS & PRACTICE FOR AS LEVEL

Maxinterax

Published by Maxinterax Ltd

12 Long Lane, Littlemore, Oxford OX4 3TW

© Max Parsonage 2008

The moral rights of the author have been asserted

First published 2008

All rights reserved. No part of this publication may be reproduced, stored in a retrieval system, or transmitted, in any form or by any means, without the prior permission in writing of Max Parsonage or as expressly permitted by law, or under terms agreed with the appropriate reprographics rights organization. Enquiries concerning reproduction outside the scope of the above should be sent to Max Parsonage, at the address above

You must not circulate this book in any other binding or cover and you must impose this same condition on any acquirer

British Library Cataloguing in Publication Data

Data available

ISBN 978-0-9555451-1-5

Originally Typeset by Magnet Harlequin

Edited and typset at Maxinterax Ltd

Printed in England by Lightning Source

Author's Acknowledgements

I am pleased to recognise the special contributions of a few, amongst many others, that have made this book more accurate, clear, and accessible. My thanks go to my wife Jane for her checking and encouragement, Dr Steve Field for checking the chemical accuracy and question answers, and Peter Mellett for ensuring clear text and precise chemistry.

The cover graphics are used with the permisson of the artist Jane Parsonage who owns the rights to the artwork.

I am grateful for the comments and enthusiasm of many students, including Toby and Hannah, as well as Sarah Salmon.

INTRODUCTION

My aim in writing this book was to make chemistry both concise and clear, so that students can quickly reinforce class work, and subsequently revise it. Through my role as an A Level Chemistry tutor I have been able to test successive drafts of this material with students, and this has been extremely valuable and interesting. It was a huge challenge to condense the essence of the subject into so few pages, but I know from working with students the value of doing this. There is just as much emphasis in the book on questions as there is on content, as only by testing themselves do students really discover how good their understanding of a topic is.

Exam-style questions are provided for students to practise their exam technique. A variety of styles is presented to reflect the diversity of questions in the papers. Students are given realistic space to write their answers, as in real exam questions. Students and teachers will be able to diagnose any exam technique problems by looking over the written answers.

MAX PARSONAGE

Max Parsonage is Head of Chemistry at d'Overbroeck's, an independent school in Oxford. He also writes about science for children, and develops interactive educational science software. He has recently produced educational software for the European Space Agency, and for Oxford University Press. In addition, he has authored and presented science vidoes.

CONTENTS

Why study chemistry?		vi
How to use this book		vii
How to succeed in your studies		vii
How to succeed in exams		viii
The A level system		ix
Exam specifications and modules		ix
1	Bonds and structures	2
2	Mass spectrometer and shape	8
3	Groups 1 and 2	14
4	Group 7 and redox	20
5	Periodic table	26
6	Organic bonding and isomers	32
7	Organic mechanisms	38
8	Alkanes and alkenes	44
9	Haloalkanes	50
10	Alcohols	56
11	Energetics: enthalpy change	62
12	Kinetics: rate	68
13	Equilibrium	74
14	Industrial processes, catalysts, and the environment	80
15	Experimental skills	86
16	Chemical calculations I	92
17	Chemical calculations I	98
Answers		104
Index		114
Periodic table		120

WHY STUDY CHEMISTRY?

Chemistry is fundamental to understanding the world around us, simply because everything is made of chemicals. From planets to cosmetics, and microbes to bridges, chemistry underpins how materials behave. It also explains how different substances can be made.

If you study AS or A level chemistry then you should be able to ask 'Why?' and receive satisfying explanations. You will find AS chemistry explains chemical ideas mostly using words, while A2 chemistry explains chemical ideas using maths as well as words.

If you like logic problems, the way ideas can just 'click' beautifully together, then you will enjoy chemistry. Once you have gained a good grasp of the chemical patterns you will find there is very little detail to memorise, because studying chemistry is like studying a game. Once you know the 'rules of chemistry', you can 'play' with the chemical ideas. Chemistry is therefore a concise subject. It is attractive because it makes you think, without requiring you to write many essays or memorise huge amounts of information. Studying chemistry complements A levels that are essay based or require a huge reading load.

You may have to take chemistry if you want to become a doctor, or vet, or if you want to study chemistry, pharmacology, environmental science, or related subjects. Chemistry is also a useful foundation for biochemistry, geology, physical geography, engineering, or materials science at university.

Because it is such a fundamental study, chemistry provides helpful background for a great variety of subjects, such as biology, pyrotechnics in theatre studies, and art restoration. If law interests you, then chemistry is a useful discipline because it encourages logical thinking.

HOW TO USE THIS BOOK

This book will supplement a standard A level textbook, or you could use it as a free-standing A level book to consult alongside your notes. It is best used as a course companion, to be referred to when starting a topic, and to help you understand when you are in difficulties. When tests and exams approach it will usefully explain things in a few pages, and test you thoroughly.

Teachers may set the book as homework, or use it in class tests.

HOW TO SUCCEED IN YOUR STUDIES

To succeed in A level chemistry you, the student, need to **understand** the ideas, **remember** the facts and ideas, and have a good **exam technique**.

An understanding of chemistry builds up layer upon layer, so the units are laid out with the foundation topics first. The units are presented in the best order for you to study them. Unit 1 (Bonds and structures) will help you understand many other topics. Similarly the early topics will help with the later ones when studying Energetics, Equilibrium, Kinetics and Organic chemistry. Understanding groups 1, 2, and 7 will help you with transition metals.

Cover one topic at a time to gain a full understanding, rather than scanning many topics quickly.

To **understand** the topics, read the relevant section. If you do not immediately understand an explanation, pause and re-read it. Use the examples to help you see the point. By writing short answers to the recall questions, test yourself to check you understand the ideas, and then refer to the answers.

To **remember** the facts and ideas, use the factual recall questions (the 'Recall test'). If you can answer all the questions in a section, then you have the facts you need to answer the exam questions on that topic (the 'Concept test'). These questions are a more effective way of memorising information than simply copying notes. You could use them to help you to identify your weaknesses, then return to the unit itself to turn your weaker topics into strengths.

HOW TO SUCCEED IN EXAMS

When you understand the concepts and have memorised any necessary ideas, you can work on improving your **exam technique**.

To gain marks in exams you must, of course, understand and know the topics. You must also have an effective exam technique; remember to read the questions carefully, make sure you understand what the examiners are asking, answer the question (rather than just writing anything you know), and communicate a clear answer using technical words correctly.

Students commonly lose marks by not reading the questions. It may appear obvious, but in the stress of the exam many students do not always read every word and so do not answer a question appropriately. So always read the question at least twice. Many students write everything they know about the subject mentioned in the question, and so produce long-winded answers. This may waste so much time that they do not finish the exam paper. More importantly, examiners state that long rambling answers tend not to gain full marks because they do not focus on the particular point raised in the question.

The examiners use particular 'command words' which indicate how you should respond to questions; marks are easily lost by ignoring these. For example, many students 'explain' when the question says 'describe'. Here is a list of command words found in A-level chemistry exam papers.

Define or What is meant by... – Give a definition in words and in equations if possible.

Describe – This asks you to state what is observed in an experiment, or state the basic points in a practical method. Giving a chemical explanation is not necessary.

Describe what will be observed – 'Observed' means seen, or sensed, so describe colours, states, and smells. Chemical names may not gain you marks.

Explain – say how and why something happens. Be careful to use the correct technical words. If many marks are offered then explain in depth.

Calculate – Obviously, work out a numerical answer. Remember to give the correct sign (for example, exothermic enthalpies are negative), and give your answer to the correct number of significant figures.

Using the data given – You must refer to the data given! Show the examiner you have done so by marking graphs, using figures in calculations, or using words from the question. Be wary of basing your answer on recall of knowledge.

Give the formula – You must give the formula, not the name. Easily overlooked.

Name – Give the name. The examiner is checking that you can name compounds. A formula will not do.

Identify – Give the chemical name or formula.

Suggest – Anything reasonable will do as an answer. There are many possible responses. This confuses some students because it is so unusual in chemistry exams; the examiner usually wants a particular answer.

Comment on – The examiner wants you to point out an idea, usually in the specification (syllabus), suggested by the data.

Here are common technical words that students misuse. Take care not to confuse them. Examiners will not give marks if you talk about atoms in sodium chloride – because it contains ions, of course.

Words for particles are atom, ion, and molecule.

Chemical substances are elements, compounds, or mixtures.

Bonding must be covalent, ionic, or metallic...

...Whereas structures must be simple covalent molecules or giant lattices (which could be covalent ionic, or metallic).

THE A LEVEL SYSTEM

The system is designed so that a typical student will study four or five ASs in their first year of study, and then select three from these to continue in their second year of study, called A2. This will then give them three full A levels. The system is designed to be flexible, so that it is possible to do any number of ASs and A2s, but an A2 can only be done if the relevant AS has been completed. In reality, most students' choices will be limited by whatever system and options their school or college can offer.

Two of the three main examination boards are offering two specifications (syllabuses) in chemistry, and one main board is offering one specification. Each AS component contains three modules, as does each A2, but the content of each module varies from specification to specification. Whatever specification a student studies, most of the content covered is the same as in any other specification, but the topics are mixed differently in the different modules.

EXAM BOARD SPECIFICATIONS AND MODULES

Module	1 AS	2 AS	3 AS
Exam Board			
AQA	1, 2, 5, 6, 7, 8, 16, 17	3, 4, 8, 9, 10, 11, 12, 13, 14	15 practical component
CCEA	1, 2, 4, 5, 16, 17	3, 6, 7, 8, 9, 10, 11, 12, 13, 14, 15	15 practical component
EDEXCEL	1, 5, 6, 7, 8, 11, 16, 17	1, 2, 3, 4, 9, 10, 12, 13, 14	15 practical component
EDEXCEL (International)	1, 5, 6, 7, 8, 11, 16, 17	1, 2, 3, 4, 9, 10, 12, 13, 14	15 practical component
OCR A	1, 2, 3, 4, 5, 16, 17	6, 7, 8, 9, 10, 11, 12, 13, 14	15 practical component
OCR (Salters)	1, 2, 3, 4, 6, 7, 8, 11, 12, 14, 16, 17	4, 5, 8, 9, 10, 12, 13, 14, 15, 16, 17	15 practical component
WJEC	1, 2, 3, 11, 12, 13, 16, 17	1, 2, 3, 4, 5, 6, 7, 8, 9, 10, 14	15 practical component

The relevant units in this book are shown for each module of each specification.

The specifications may be downloaded from the relevant exam board. They are useful when revising.

Unit 1: BONDS AND STRUCTURES

Every substance is made of atoms. The electronic structure determines the types of bonds between the atoms. In turn the arrangement of atoms and the bonding between them determines the physical and chemical properties of all substances.

- Atoms have structure. The central **nucleus** contains positively charged protons (+) and neutral neutrons (0) and is surrounded by layers, called **shells**, of negatively charged electrons (–). The electron number is equal to the proton number (atomic number) in a neutral (uncharged) atom. All the atoms in a given element have the same proton number. Atoms of different elements react differently, due to the actual electron number in their shells. See Fig. 1.1.

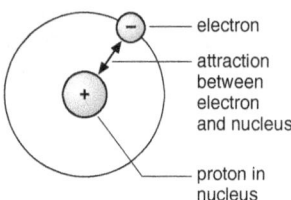

Fig. 1.1

- Elements are arranged in the **periodic table** in order of increasing proton number. The proton number increases from left to right across horizontal rows (called **periods**) and from the top to the bottom of vertical columns (called **groups**). Look at the periodic table printed in this book as you study the next section.

  Across a period, increasing proton number increases the attraction of the nucleus for the electrons in the outer shell. Therefore, **atomic radius** decreases as proton number increases across a period.

  Down a group, the number of electron shells increases as proton number increases. Filled inner shells increase the distance of the outer shell from the nucleus. These filled shells also **shield** electrons in the outer shell from the attraction of the full nuclear charge. Therefore, atomic radius increases as proton number increases down a group.

- **Electronegativity** is a measure of how attractive an atom is for a pair of electrons in a covalent bond. An element's attracting ability increases with increasing electronegativity value. See Fig. 1.2 and Fig. 1.3.

  Electronegativity values affect the type of bond that forms.

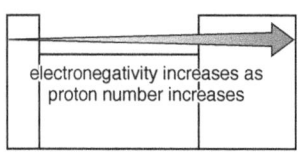

Fig. 1.2

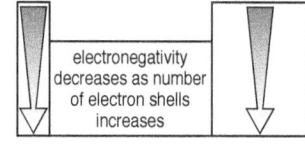

Fig. 1.3

Fig. 1.5

CATions are PUSSYtive

Ionic compounds generally dissolve in water because water molecules are polar.

Water molecules are polar because they contain polar covalent bonds. See page 4.

- **Ionic bonds** occur when one atom is much more electronegative than another atom. The atom with the smaller electronegativity loses an electron to the atom with the greater electronegativity. The electron acceptor gains (–) electrons and so becomes a negative ion (**anion**) while the electron donor loses electrons to become a positive ion (**cation**). See Fig. 1.4 and Fig. 1.5.

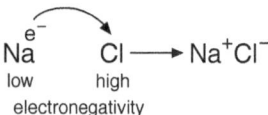

Fig. 1.4

Anions and cations combine in a regular pattern to form **giant ionic lattices**. Melting points are high due to the strong electrostatic forces between the oppositely charged ions. See Fig. 1.6.

Ionic bond strength is greatest when the ions have large charges and small ionic radii.

Ionic compounds are **brittle**. External forces displace ions so that similar charges are next to each other. They repel and the lattice breaks.

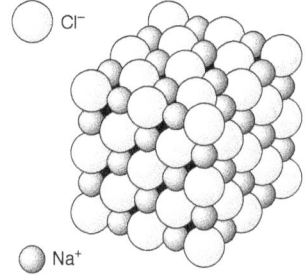

Fig. 1.6

- **Metals** have relatively small electronegativity values. **Metallic bonds** form when weakly held electrons in the outer shell become mobile and move freely between atoms. The resulting metal cations are bonded together by their attraction for the mobile (**delocalised**) electrons. They arrange in regular patterns to form **giant metallic lattices**. See Fig. 1.7.

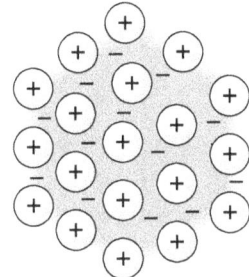

Fig. 1.7

Metallic bond strength, and hence hardness and melting point, generally increase with increasing numbers of delocalised electrons. Smaller atoms make for stronger structures.

- **Non-metals** have relatively large electronegativity values. Atoms form **covalent bonds** by sharing electrons (see Fig. 1.8). Covalent bond strength generally increases with increasing electronegativity. **Dative covalent (co-ordinate) bonds** occur when both electrons in a covalent bond come from only one of the atoms.

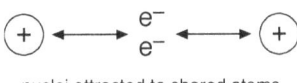

nuclei attracted to shared atoms

Fig. 1.8 A covalent bond

- There are two sorts of covalent structures. **Giant covalent (macromolecular) lattices** (e.g. diamond) tend to be very hard because each atom has a strong covalent bond to the next atom in the structure. **Simple covalent molecules** (e.g. water and chlorine) have strong covalent bonds between the atoms that make up each molecule, but there are only weak forces of attraction between the molecules. See Fig. 1.9.

- Ionic bonding and covalent bonding represent two extreme forms of **bonding character**. Ionic compounds have a degree of covalent character due to incomplete transfer of electrons. Covalent compounds have a degree of ionic character due to unequal sharing of electrons.

All the giant lattices (metallic, ionic, and covalent) have high **melting points** and **boiling points** because large amounts of energy are needed to break the bonds and separate the atoms or ions.

Unit 1 — INTERMOLECULAR FORCES

- There are three types of **intermolecular forces** of attraction between simple covalent molecules; Van der Waals forces, permanent dipole interactions, and hydrogen bonding. Molecules with hydrogen bonding are more attractive than those with just a permanent dipole, which are more attractive than those with just Van der Waals forces.

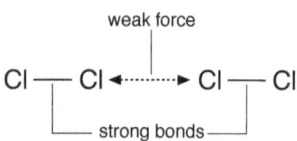

Fig. 1.9

Relative molecular mass (RMM) is sometimes used instead of M_r.

Weak **Van der Waals forces** (which you may know as fluctuating dipole, or London forces) exist between all molecules. These forces (and so the melting and boiling points) increase with increasing numbers of electrons and molecular length, e.g. methane CH_4 ($M_r = 16$) b.p. (boiling point) = $-161\cdot°C$; straight-chain butane C_4H_{10} ($M_r = 58$) b.p. = $-0.5\cdot°C$; globular methylpropane C_4H_{10} ($M_r = 58$) b.p. = $-11.7\cdot°C$. See Fig. 1.10. (M_r is the **relative formula mass**, calculated by adding up the **relative atomic masses** of the elements in the formula.)

Permanent dipole interactions exist between molecules that have **polar covalent bonds**. This type of intermolecular force results when atoms in a covalent bond have different electronegativities so that bonding electrons are shared unequally between them. **Example:** In H-Cl the Cl atom is much more electronegative than the H atom so the covalent bond is polar $^{\delta+}$H-Cl$^{\delta-}$ (see Fig. 1.11). NB In CCl_4, all the C-Cl bonds are polar, but the molecule is symmetrical, making the overall molecule non-polar. You will mainly meet permanent dipoles in covalent molecules that contain halogens or C=O or C-O bonds.

If you compare different molecules of similar M_r, those with polar bonds (and hence permanent dipole interactions) have higher melting points than those with Van der Waals forces only.

Hydrogen bonds form when two conditions are satisfied:
(1) a hydrogen atom is covalently bonded to a highly electronegative atom, so that it becomes electron deficient;
(2) a small strongly electronegative atom with a **lone pair** (see left) is present e.g. N, O, or F atoms. (They spell NOF).

A **lone pair** is a non-bonding electron pair in an atom's outer shell, drawn like this: ⊙

Hydrogen bonds form between the negative charge associated with the lone pair of electrons and the δ+ H atom. **Example:** Ethanol (see Fig. 1.12) and water have H-O bonds so the H atom is electron deficient; the O atom has lone pairs of electrons. By contrast, all the H atoms in ethanal CH_3CHO are joined to C atoms so none is electron deficient. Hydrogen bonding is not present in ethanal, although there are permanent dipole interactions between $^{\delta+}$C=O$^{\delta-}$ bonds. So ethanol has a higher boiling point than ethanal.

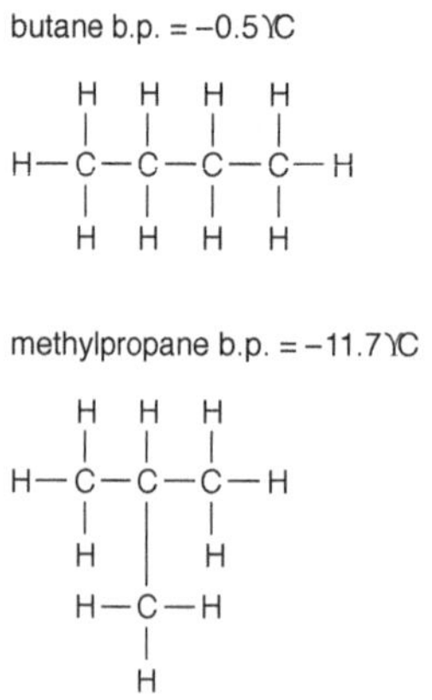

Fig. 1.10 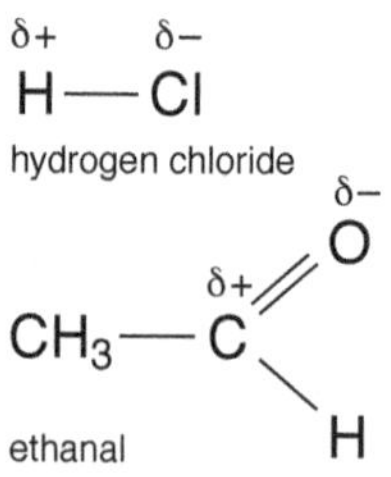   Fig. 1.11 Polar molecules 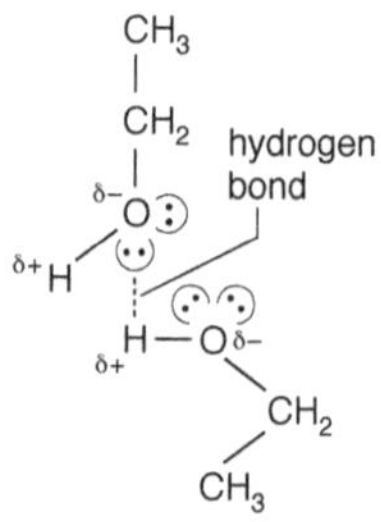   Fig. 1.12 Ethanol H-bonds

TESTS
RECALL TEST

1. What determines the chemical properties of an element?

   _____ (1)

2. State and explain the trend in electronegativity across the third row of the periodic table.

   _____ (2)

3. State and explain the trend in electronegativity down the second group of the periodic table.

   _____ (3)

4. Why is AgI covalent?

   _____ (1)

5. Why are some bonds polar?

   _____ (1)

6. What are the three intermolecular forces?

   _____ (1)

7. If the covalent bonds are so strong why is chlorine a gas?

   _____ (1)

8. If the covalent bonds in I_2 are weak why is iodine a solid?

   _____ (1)

9. Why does chloromethane have a dipole?

   _____ (1)

10. What is necessary for hydrogen bonding to occur?

    _____ (4)

11. Why does ethanol have a higher boiling point than ethanal?

    _____ (2)

12. Why does NaCl conduct when molten, but not when solid?

    _____ (2)

Unit 1 TESTS

13 Why do the boiling points of the noble gases increase with increasing atomic number?

_____

_____ (1)

14 Why does water have a higher boiling point than the rest of the group-6-hydrides?

_____

_____ (1)

15 When chlorine gas is cooled, why does it condense into a liquid?

_____

_____

_____

_____ (4)

16 When solid NaCl is heated from room temperature it melts. Explain the change.

_____

_____

_____

_____ (4)

(Total 30 marks)

CONCEPT TEST

1 a Define the term 'electronegativity'.

_____

_____ (2)

b State and explain the bonding in carbon dioxide.

_____

_____ (2)

c Glucose molecules have many -OH groups. Why is glucose soluble in water?

_____

_____

_____ (2)

d One explanation for the covalent character of beryllium chloride is that the covalent bonds are due to the Be^{2+} cation polarising the Cl^- anion so that the electrons in the anion are shared by the cation).

  i Why does Be^{2+} polarise Cl^- when Ca^{2+} cannot?

_____

_____

_____ (1)

ii Why is aluminium chloride covalent?

_____

_____

_____
(2)

iii State and explain the difference in size between a chloride ion (Cl^-) and a fluoride ion (F^-).

_____
(2)

iv Why is aluminium fluoride ionic?

_____

_____
(1)

2 For many years people have known two types of carbon: diamond and graphite. A third form was discovered recently containing C_{60} molecules, called buckminsterfullerene, or 'bucky balls' for short. It contains carbon atoms in a ball, a bit like a modern football, made of hexagons and pentagons. In the molecule of C_{60} the carbon atoms would be where three of the shapes join.

a Explain why graphite is soft.

_____
(2)

b Explain why diamond is strong.

_____

_____
(2)

c Knowing C_{60} is made of balls, predict the properties of C_{60} at room temperature and pressure.

_____

_____

_____
(2)

d One idea is to trap K^+ ions inside the C_{60} to make KC_{60}^+ ions. These cations could be combined with Cl^- ions. State and explain the magnitude of boiling point that this substance would have.

_____

_____
(2)

(Total 20 marks)

Unit 2 — SHAPE AND MASS SPECTROMETER

- **Molecular shape** can be decided by using **VSEPR** (**valence shell electron pair repulsion**) theory, which depends on the number of **bonding** pairs and **lone** (non-bonding) pairs of electrons surrounding the central atom.

- Firstly, identify bonding and lone electron pairs by drawing a **dot and cross** diagram showing the **outer** (**valence shell**) **electrons** of each atom. Remember that negative ions have one or more extra electrons and that positive ions lack one or more electrons.

  Molecular shape results from the electron pairs **repelling** each other so that they are as **far apart** as possible. You should expect to explain this every time an exam question discusses shape.

  Lone pairs repel more than bonding pairs, so lone pairs tend to push the bonding pairs together. Repulsion decreases in the order: lone pair–lone pair > lone pair–bond pair > bond pair–bond pair.

  Double bonds behave as single bonds, but with increased electron density and therefore an increased repulsive effect. For examples of molecular shapes, see Fig. 2.1.

Fig. 2.1

Formula	Pairs of electrons	Bonding pairs	Lone pairs	Dot and cross diagram	Drawn shape	Name of shape	Similarly shaped ions
$BeCl_2$	2	2	0	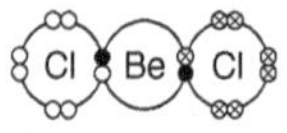	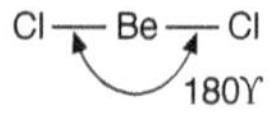 Cl—Be—Cl 180°	linear	
BF_3	3	3	0	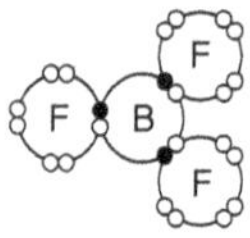	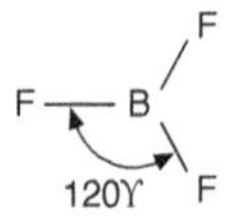 F—B with F, F at 120°	trigonal planar	NO_3^-, CO_3^{2-}
CH_4	4	4	0	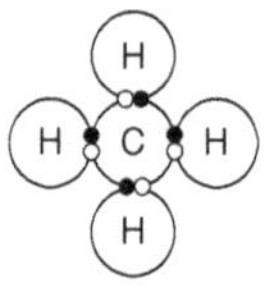	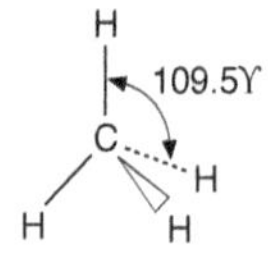 109.5°	tetrahedral	PCl_4^+, SO_4^{2-}, NH_4^+
NH_3	4	3	1	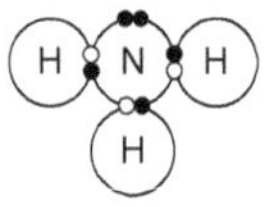	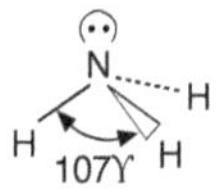 107°	pyramidal	SO_3^{2-}
H_2O	4	2	2	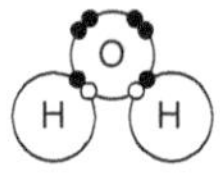	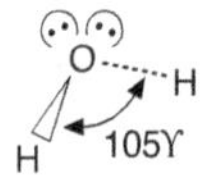 105°	V-shaped	
PCl_5	5	5	0	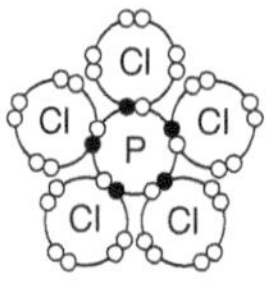	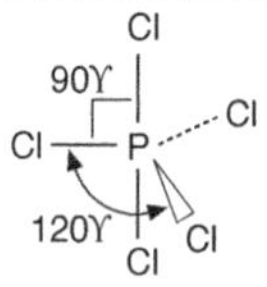 90°, 120°	trigonal bipyramidal	
SF_6	6	6	0	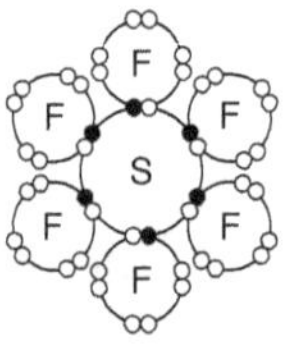	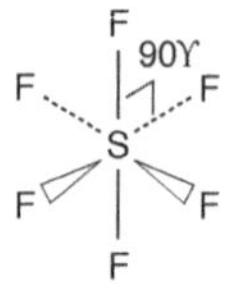 90°	octahedral	PCl_6^-

shape of ethene: ~120°, $H_2C=CH_2$

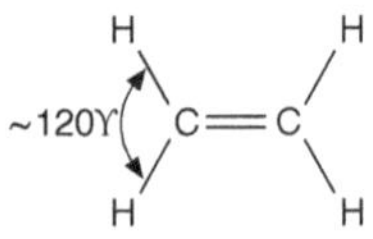

shape of SO_2

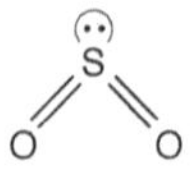

Fig. 2.2

Note these other examples:

carbon dioxide CO_2 is **linear**;

ethene CH_2CH_2 is **trigonal planar** around each carbon atom;

SO_2 is **V-shaped** because of the presence of a lone pair (see Fig. 2.2).

- Always carry out the steps shown in the following example:

  Question: What is the shape of XeF$_4$?
  (i) The number of electrons in the outer shell of the Xe atom is eight; there are seven outer-shell electrons in each F atom.
  (ii) Each F atom shares one electron with the Xe atom to make one covalent bond.
  (iii) Four Xe electrons are used to bond with the four F atoms.
  (iv) Four of the Xe electrons are not used for bonding and exist as two lone pairs.
  (v) There are six electron pairs around the Xe atom, so the shape of XeF$_4$ is based on an octahedron. Lone pairs are opposite each other, not adjacent (see Fig. 2.3) so the actual shape is square planar.

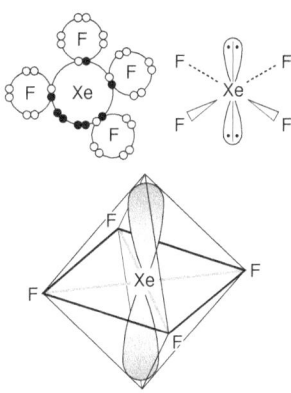

Fig. 2.3

- There are six main points for you to remember about the functioning of a **mass spectrometer** (see Fig. 2.4).
  1. The sample is **vaporised** – to make the molecules or atoms mobile.
  2. High-energy electrons bombard the molecules or atoms in the vapour. Electrons are ejected and **cations form**.
  3. The cations are **accelerated** and **focused** into a beam by **electric fields**.
  4. The cation beam is **deflected** by a **magnetic field**. The angle of deflection increases with decreasing cation mass. The magnetic field strength is varied so that ions of known mass-to-charge ratio enter the **detector**.
  5. Ions entering the detector cause a **current** to flow in an external circuit connected to a **recording device**.
  6. A pump maintains a **vacuum** inside the apparatus. Otherwise the cations would be scattered by air molecules.

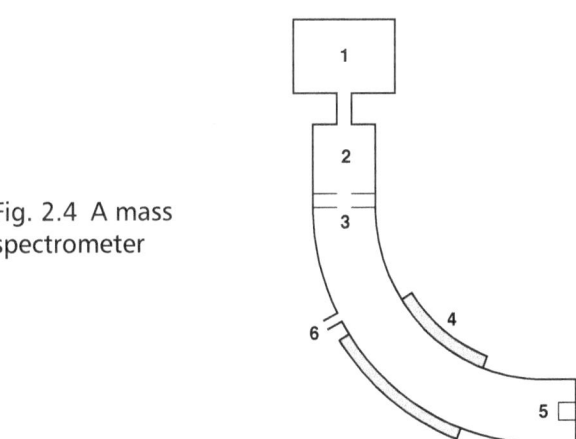

Fig. 2.4 A mass spectrometer

- You must remember these definitions:

  Atomic number (Z): The number of protons in the nucleus of an atom.

  Mass number (A): The total number of protons and neutrons in the nucleus.

  Isotopes: Atoms of the same element with different mass numbers (i.e. isotopes have the same atomic numbers but different mass numbers).

  Relative atomic mass: The weighted average mass of the atoms in a sample of an element divided by $\frac{1}{12}$th of the mass of an atom of the carbon-12 (^{12}C) nuclide.

  Relative isotopic mass: The mass of the atoms in a sample of an isotope divided by $\frac{1}{12}$th of the mass of an atom of the carbon-12 (^{12}C) nuclide.

Unit 2 — EXAMPLES

You must be able to calculate relative atomic mass from **isotopic mass** and **relative abundance** data (usually obtained from a mass spectrometer).

Lay out your calculation the same way each time and then you will make fewer mistakes. **Example:**

Isotope	Abundance
^{63}Cu	69.1%
^{65}Cu	30.9%

Mass of 100 atoms = $(63 \times 69.1) + (65 \times 30.9) = 6361.8$

Average mass of 1 atom = $\dfrac{\text{total mass of the atoms}}{\text{the number of those atoms}} = \dfrac{6361.8}{100}$

= 63.6 (to 3 significant figures)

You must add up the abundances to find the total number of atoms, not the masses. The total abundance may not be 100, especially if you have to measure the value from a graph (called a '**mass spectrum**'). **Example:** See Fig. 2.5.

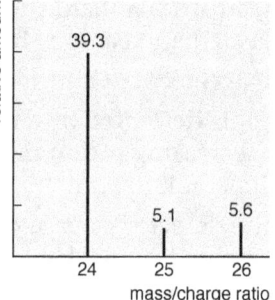

Fig. 2.5 Graph of the mass spectrum of magnesium

Calculation:

$24 \times 39.3 = 943.2$
$25 \times 5.1 = 127.5$
$26 \times 5.6 = 145.6$
$\overline{50 \quad\quad 1216.3}$

$1216.3 \div 50 = 24.326$

relative atomic mass (RAM) = 24.3

You may see a mass spectrum of a compound that consists of **molecules**. Often, some of the original molecules are detected as well as fragments of that molecule. If the original molecule is detected, it is called the '**molecular ion**' or '**parent ion**'. The molecular ion will have the largest mass so, usually, the greatest mass (not the greatest abundance) on the spectrum will indicate the relative molecular mass of the molecule.

Example: See Fig. 2.6.

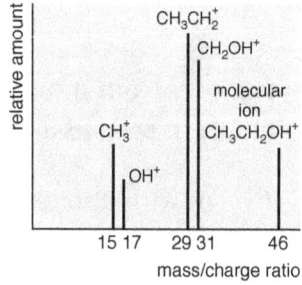

Fig. 2.6 Mass spectrum of ethanol (CH_3CH_2OH)

TESTS
RECALL TEST

1. Name the particles in the nucleus and give their charges.

   _____ (2)

2. State the processes that take place in the mass spectrometer.

   _____ (6)

3. Define

   a isotopes

   _____ (2)

   b relative atomic mass

   _____ (2)

   c mass number

   _____ (2)

4. In a mass spectrum, which peak usually indicates the relative molecular mass of a substance: the peak on the left, the tallest peak, or the peak on the right? (1)

5. Name the shapes of these substances:

   water _____, sulfur hexafluoride _____,

   methane _____, beryllium chloride _____,

   boron trifluoride _____, ammonia _____,

   phosphorus pentachloride _____. (7)

6. Work out the shape of the ammonium ion (NH_4^+). What must the bond angle be? (2)

7. What is the shape of the hydroxonium ion, H_3O^+? (1)

8. State the bond angles in: methane _____, ammonia _____,

   water _____, carbon dioxide _____. (4)

9. What must you state (almost) every time you discuss shape of molecules in-exams?

   _____ (1)

   (Total 30 marks)

Unit 2 — TESTS

CONCEPT TEST

1 a In the mass spectrometer:

  i How are the ions separated according to mass?

  _____ (1)

  ii Why is a vacuum pump used?

  _____ (1)

Abundance	Isotopic mass
25.7	107
24.3	109

  iii Calculate the relative atomic mass of silver from the data left.

  _____ (2)

b A pure sample of P, a possible pollutant, is put through a mass spectrometer. Chemical analysis indicates the *empirical formula* of the substance is CH_2O. There are peaks on the mass spectrum at 15, 28, 45, 60.

  i What is the formula mass of P?

  _____ (1)

  ii Which particles produced the peaks with these mass/charge ratios?

  15 _____

  28 _____

  45 _____

  60 _____ (4)

  iii Using the formula mass and the empirical formula, give two possible structures for P. (2)

  iv Using the information in **ii**, state the structure of P. Explain your answer.

  _____
  _____
  _____ (2)

2 The mass spectrum of nickel (atomic number 28) produces the data in the margin:

Abundance	Mass number
33.95	58
13.1	60
0.6	61
1.85	62
0.5	64

  a Name the particles in the nucleus and state how many of each are found in the Ni-58 nucleus.

  _____
  _____
  _____
  _____ (3)

  b Calculate the relative atomic mass of nickel.

  _____
  _____
  _____
  _____ (2)

3 a Explain why $AlCl_3$ is trigonal planar while NH_3 is pyramidal.

  _____
  _____
  _____
  _____
  _____ (3)

  b Draw the shape of these molecules and ions:

  i PH_3 **ii** SO_2 **iii** ClO_3^- **iv** BrF_3^- (4)

  c Name the shape of SF_6.

  _____ (1)

  d State and explain the different bond angles in CH_4, NH_3, and H_2O.

  Bond angles:

  CH_4 _____

  NH_3 _____

  H_2O _____

  Explanation for differences:

  _____
  _____
  _____ (4)

  (Total 30 marks)

Unit 3: GROUPS 1 AND 2

- The elements of groups 1 and 2 make up the **s-block** elements. With increasing atomic number in each group, **metallic bond strength decreases** and the metals become **softer** and have **lower melting points**. Outer bonding electrons become further from the nucleus and are **shielded** from its charge by filled inner shells.

- Similarly, **electronegativity** and **1st ionisation energies** (see unit 5) **decrease** as the electrons in the outer shell become further from the nucleus (see Fig. 3.1).

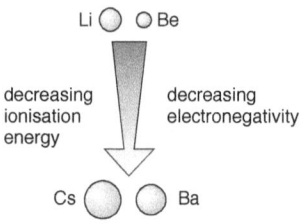

Fig. 3.1

For a given period, the group 2 element has **one more proton** in its nucleus and **one more electron** in its outer shell than the group 1 element. Therefore, group 2 metallic bonds are **stronger** than those of group 1 and group 2 metals are harder and have **higher melting points**.

Similarly, the **electronegativities** and **1st ionisation energies** of the group-2 elements are **higher** than those of the corresponding group 1 elements because the group 2 nuclei have a **greater attraction** for the outer electrons (see Fig. 3.2).

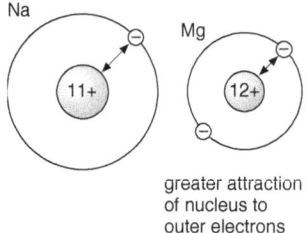

Fig. 3.2

- Group 1 atoms always **lose** control of one electron when they bond: the **oxidation number** in their compounds is always +1. Similarly, the group-2 oxidation number is always +2. (See the section on group-7 for a full explanation of oxidation number).

- All s-block compounds are **predominantly ionic** except $BeCl_2$ (see unit-1 for further explanation).

- **Lithium** and **beryllium** have many properties that are **unlike** those of other members of their groups, e.g. lithium is the only group 1 metal that reacts with nitrogen to form a compound; molten beryllium compounds are poor electrical conductors, indicating strongly covalent character.

- S-block metal compounds usually produce **coloured flames** when strongly heated. Heat **promotes electrons** from lower **energy levels** to higher ones. When the electrons return to lower energy levels, characteristic coloured visible light is emitted, because the energy-level difference is the same as the energy of visible light (see Fig. 3.3).

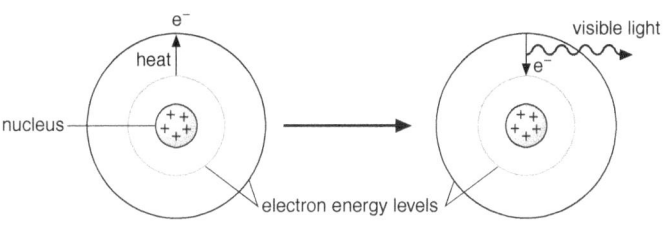

Fig. 3.3

Flame colours

- **Li** red (crimson)
- **Na** yellow
- **Ca** orange-red
- **Sr** red (crimson)
- **Ba** pale green (apple green)
- **K** lilac or pale purple*
- **Mg** compounds **do not** produce a coloured flame

*You should state that the lilac colour is visible through blue glass.

- All s-block elements (except Be) react with chlorine to form **ionic chlorides**. Examples:
  Group 1: $2Na(s) + Cl_2(g) \rightarrow 2NaCl(s)$
  Group 2: $Mg(s) + Cl_2(g) \rightarrow MgCl_2(s)$

- All s-block elements react with oxygen to produce **ionic oxides**, e.g.
  Group 1: $4Li(s) + O_2(g) \rightarrow 2Li_2O(s)$
  also $2Na(s) + O_2(g) \rightarrow Na_2O_2(s)$ as well as Na_2O (see opposite)
  Group 2: $2Mg(s) + O_2(g) \rightarrow 2MgO(s)$
  NB **BeO is amphoteric** (like Al_2O_3) showing that Be has non-metallic character.

Amphoteric substances react with both acids and bases.

- In general, s-block elements (not Be) **react with water** to produce hydrogen gas and a metal hydroxide, though some form an insoluble metal oxide (e.g. MgO).
  Group 1: $2Na(s) + 2H_2O(l) \rightarrow 2NaOH(aq) + H_2(g)$
  Group 2: $Ca(s) + 2H_2O(l) \rightarrow Ca(OH)_2(aq) + H_2(g)$
  in **steam**, $Mg(s) + H_2O(g) \rightarrow MgO(s) + H_2(g)$

- You must recall the **acid–base reactions** (even though your speciifcation may not state this).
  Metal + Acid → Salt + Hydrogen
  e.g. $Mg(s) + H_2SO_4(aq) \rightarrow MgSO_4(aq) + H_2(g)$
  Metal oxide + Acid → Salt + Water
  e.g. $MgO(s) + H_2SO_4(aq) \rightarrow MgSO_4(aq) + H_2O(l)$
  Metal hydroxide + Acid → Salt + Water
  e.g. $Mg(OH)_2(s) + H_2SO_4(aq) \rightarrow MgSO_4(aq) + 2H_2O(l)$
  Metal carbonate + Acid → Salt + Water + Carbon dioxide
  e.g. $MgCO_3(s) + H_2SO_4(aq) \rightarrow MgSO_4(aq) + H_2O(l) + CO_2(g)$

- Reaction of group 1 metals with **oxygen** can form **simple oxides** (O^{2-} ions), **peroxides** (O_2^{2-} ions), and **superoxides** (O_2^- ions). You may not meet these in your studies, but sometimes examiners include them in exam questions. Peroxides and superoxides are destabilised by small cations; stability increases with increasing cation size. See right.

The Group 1 oxides

Li_2O

Na_2O, Na_2O_2

K_2O, K_2O_2, KO_2

(Rb and Cs as K)

Unit 3: THERMAL DECOMPOSITION & SOLUBILITY

- All **group 2 carbonates** decompose when heated to form the oxide and carbon dioxide, e.g. $CaCO_3(s) \rightarrow CaO(s) + CO_2(g)$

  Examine the table below. It shows the temperature at which the group-2 carbonates decompose and the size of the ions.

Fig. 3.4 Table of carbonate decomposition temperature and ion size

Compound	Decomposition temperature (K)	Cation radius (10^{-1} nm)
$BeCO_3$	370	0.31
$MgCO_3$	810	0.65
$CaCO_3$	1170	0.99
$SrCO_3$	1560	1.13
$BaCO_3$	1630	1.35

The **decomposition temperature increases** as the cation size increases. Carbonate stability decreases with **increasing polarisation** by the cation. The cation distorts the electron shells of the anion and draws them towards itself, **increasing covalent character** (see Fig. 3.5).

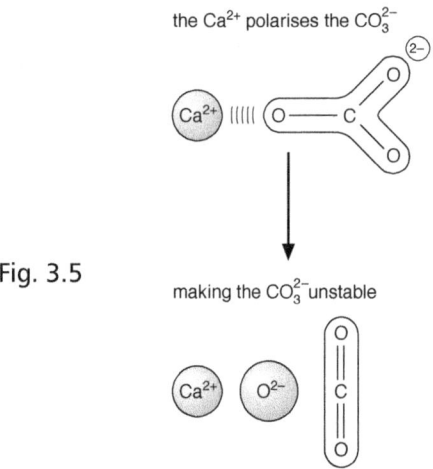

Fig. 3.5

- In general, covalent character is more likely if:
  the charge on the ions is high; AND the cation is small or the anion is large.

  Group 1 carbonates generally do not decompose because the cation charge is only 1+ (and the cations are large), so does not polarise the anion sufficiently. However **Li⁺ ions** are very small so they will polarise the carbonate anion:

  $Li_2CO_3(s) \rightarrow Li_2O(s) + CO_2(g)$

- Generally, an ionic solid is **insoluble** if the energy required to separate the ions from their lattice is much more than the energy released when the same ions are surrounded by water (i.e. are **hydrated**). This means that small (endothermic) **lattice energy** and large (exothermic) **hydration energy** indicate good solubility.

- The **solubilities of group 2 sulfates decrease** with increasing atomic number.

  Barium sulfate is extremely insoluble. Barium ions are used to **test** for the presence of aqueous sulfate anions. Mixing acidified barium chloride (or nitrate) solution with an aqueous sulfate solution produces a thick white precipitate. The acid reacts with sulfite or carbonate and prevents them forming white precipitates of barium sulfite or barium carbonate.

  By contrast, the **solubilities of group 2 hydroxides increase** with increasing atomic number.

TESTS
RECALL TEST

1. Explain why magnesium atoms are smaller than sodium atoms.
   _____
   _____
   _____ (2)

2. State the group 2 chlorides' flame colours: Mg _____,

   Ca _____, Sr _____, Ba _____,

   Li _____, Na _____, K _____. (7)

3. What else must you mention when stating the colour of the potassium flame?
   _____ (1)

4. Why is barium chloride more ionic than magnesium chloride?
   _____
   _____ (1)

5. Write a balanced equation for each of these reactions:

   a magnesium + hydrochloric acid _____

   b magnesium oxide + hydrochloric acid _____

   c magnesium carbonate + hydrochloric acid _____

   d magnesium + chlorine _____

   e magnesium + oxygen _____ (5)

6. Why is it difficult to form magnesium peroxide?
   _____ (2)

7. What is the trend in the solubility of the group 2 sulfates?
   _____ (1)

8. Explain why the group 2 hydroxides become more soluble with increasing atomic number.
   _____
   _____
   _____ (1)

Unit 3 — TESTS

9 What is the test for sulfate ions?

_____

_____ (2)

10 Write an equation for the reaction of limewater with carbon dioxide.

_____

_____ (1)

11 Explain why lithium nitrate is unstable to heat.

_____

_____

_____ (2)

12 Write an equation for the thermal decomposition of barium carbonate.

_____

_____ (1)

13 When sodium peroxide is heated oxygen is given off. Write an equation for this reaction.

_____

_____ (1)

14 Barium peroxide and water form hydrogen peroxide and barium hydroxide. Write an equation for this reaction.

_____

_____ (1)

(Total 30 marks)

CONCEPT TEST

1 a When sodium is heated in a Bunsen flame a characteristic yellow flame is seen. Explain why sodium compounds produce coloured flames.

_____

_____

_____ (3)

b Sodium chloride may be made by burning sodium in chlorine. Explain why caesium chloride is not usually made in this way in the laboratory.

_____

_____

_____ (2)

c Write an equation to show the action of water on sodium chloride.

_____

_____ (1)

d Sodium iodide is ionic, while lithium iodide is covalent. Explain why this is so.

_____

_____

_____ (3)

2 Group 2 elements and compounds show some marked trends in physical and chemical properties within the group, with increasing atomic number.

a State and explain the trend in solubility of the group 2 sulfates with increasing atomic number.

_____

_____

_____

_____

(4)

b Aqueous barium ions form a heavy white precipitate with a particular aqueous anion even when acid is added. What is this anion?

_____ (1)

c Explain why the group 2 fluorides increase in solubility with increasing atomic number.

_____

_____

_____ (3)

d Explain why magnesium carbonate decomposes at a moderate temperature but barium carbonate is stable until heated to much higher temperatures.

_____

_____

_____ (2)

e Apply your understanding of group 2 thermal stability to explain why aluminium carbonate decomposes more easily than magnesium carbonate.

_____

_____

_____

_____ (1)

(20 marks)

Unit 4: GROUP 7 AND REDOX

Group 7 elements are also known as **halogens** ('salt generators'); halogen salts are called **halides**.

- You must recall the **colours** and **states** (at room temperature).

Element	Formula	Colour	State (room temp.)
Fluorine	F_2	Pale yellow	gas
Chlorine	Cl_2	Green-yellow	gas
Bromine	Br_2	Red-brown	liquid
Iodine	I_2	Black	solid

Iodine is a black solid, **purple** in non-polar solvents, **brown** in polar solvents. It is readily soluble in aqueous KI and turns starch blue-black.

As you can see, **melting** and **boiling** points increase with increasing atomic number. This trend is the result of the **Van der Waals forces** increasing with the size of the molecules and the number of electrons present (see Fig. 4.1).

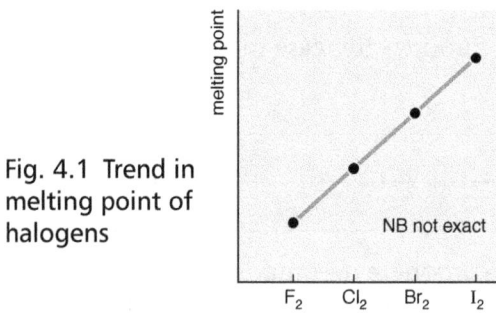

Fig. 4.1 Trend in melting point of halogens

In contrast, **bond strengths** weaken from Cl_2 to I_2 because filled inner shells increase the size of the atoms so that their nuclei are further from the shared (valence) electrons (see Fig. 4.2).

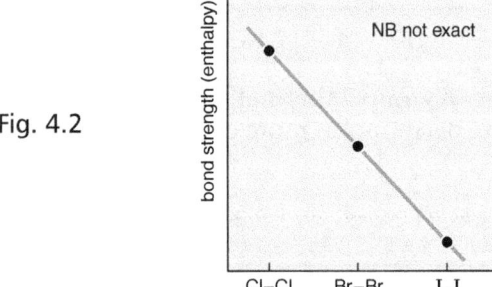

Fig. 4.2

All these elements are strongly attractive to electrons because they each have a high proton number for the period they occupy. The decrease in **electronegativity** and the decrease in **1st ionisation energy** with increasing atomic number are explained in unit 1. (See also unit 5.)

- **Halogen reactivity** decreases from F to I as the product lattice or bond energies decrease.

- Halogen/halide **displacement** reactions can happen as the elements compete for electrons. The more reactive halogen will displace a less reactive halogen from a solution containing its ions, e.g.

  $Cl_2(aq) + 2Br^-(aq) \rightarrow Br_2(aq) + 2Cl^-(aq)$

  $Br_2(aq) + 2Cl^-(aq)$ do not react because Br_2 is not a sufficiently powerful oxidant to oxidise Cl^- ions.

- Chlorine and bromine **bleach** litmus.

- Chlorine is made industrially by the **electrolysis of brine** (a concentrated solution of impure NaCl) in a membrane (or diaphragm) cell. The **membrane** keeps the electrolysis products separate and stops them from reacting together. Chlorine is produced at the (+) **anode**:

  $2Cl^-(aq) \rightarrow Cl_2(g) + 2e^-$

  The solution passes through the membrane and H_2 is produced at the (−) **cathode**:

  $2H_2O(l) + 2e^- \rightarrow 2OH^-(aq) + H_2(g)$

  Overall: $NaCl(aq) + 2H_2O(l) \rightarrow 2NaOH(aq) + H_2(g) + Cl_2(g)$.

  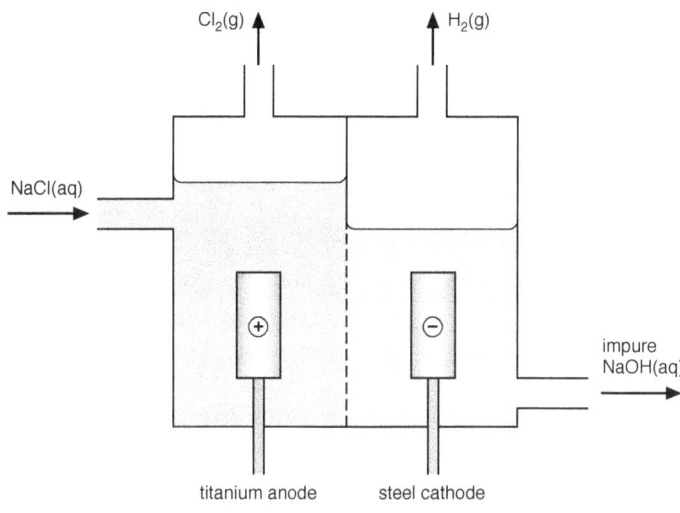

  Fig. 4.3 Electrolysis of aqueous NaCl in a diaphragm cell

  Sodium hydroxide is extracted from the brine that flows out of the cell (see Fig. 4.3).

  All the elements react directly with **metals** to form halides, e.g.

  $2Fe(s) + 3Cl_2(g) \rightarrow 2FeCl_3(s)$

- Aqueous **silver nitrate** is used as a **test** for halide ions, forming distinctive silver halide precipitates:

  $Ag^+(aq) + Cl^-(aq) \rightarrow AgCl(s)$ (white solid, soluble in dilute $NH_3(aq)$)

  $Ag^+(aq) + Br^-(aq) \rightarrow AgBr(s)$ (cream off-white solid soluble in conc. $NH_3(aq)$)

  $Ag^+(aq) + I^-(aq) \rightarrow AgI(s)$ (pale yellow solid insoluble in aqueous ammonia)

  In contrast **AgF** does not form a precipitate due to the large hydration energy of the small F^- ion.

- HCl, HBr, and HI are gases that are very soluble in water, forming strongly acidic solutions, e.g. $HCl(g) + (aq) \rightarrow H^+(aq) + Cl^-(aq)$.

- **Hydrogen chloride** is produced by mixing (non-volatile) concentrated sulfuric acid with an ionic chloride, e.g.

  $NaCl(s) + H_2SO_4(l) \rightarrow NaHSO_4(aq) + HCl(g)$ (similarly with KF)

  If kept cold KBr will produce **hydrogen bromide**; if hot, then the HBr will reduce the sulfuric acid (an oxidising acid when concentrated) to SO_2:

  $2HBr(g) + H_2SO_4(l) \rightarrow Br_2(g) + SO_2(g) + 2H_2O(l)$

  Hydrogen iodide will further reduce sulfuric acid to form I_2, S, H_2S, and H_2O.

- Chlorine dissolves in water to make **chloric(I) acid**, HClO:

  $Cl_2(g) + H_2O(l) \rightleftharpoons HCl(aq) + HClO(aq)$

  Note that the oxidation number of chlorine is 0 in the Cl_2 and in the products is −1 in HCl and +1 in the **chlorate(I)** ion ClO^-. This simultaneous oxidation and reduction of the same element in a reaction (see below) is called **disproportionation**. It is the HClO in aqueous chlorine that bleaches colour and kills bacteria in drinking water and swimming pools.

  Commercial **bleach** is made by dissolving chlorine in cold aqueous NaOH:

  $Cl_2(g) + 2NaOH(aq) \rightarrow NaCl(aq) + NaClO(aq) + H_2O(l)$

  If warmed the chlorate(I) further disproportionates to give **chlorate(V)**:

  $3ClO^-(aq) \rightarrow ClO_3^-(aq) + 2Cl^-(aq)$

  NaClO and potassium iodide produce brown iodine.

- To determine the amount of iodine in solution, sodium thiosulfate is used with starch as an indicator to help show the presence of iodine:

  $I_2(\text{in KI(aq)}) + 2S_2O_3^{2-}(aq) \rightarrow 2I^-(aq) + S_4O_6^{2-}(aq)$

Unit 4 — OXIDATION AND REDUCTION

Redox reactions involve electron transfer. **O**xidation **I**s the **L**oss of electrons; **R**eduction **I**s the **G**ain of electrons. **OILRIG**.

- An **oxidising agent** (oxidant) is a substance that oxidises another substance and is itself reduced in the process. An oxidising agent takes electrons from another substance which acts as a **reducing agent** (reductant).

- **Oxidation number** is the number of electrons an atom has gained or lost control of as a result of its bonding, e.g. when sodium atoms react, they lose one electron per atom so the oxidation number is +1 (note that charge is written the opposite way as 1+).

  Half equations are chemical equations which show the redox change for one substance, with electrons added to balance the equation, e.g. the reaction between chlorate(I) ions and iodide ions in acidified solution to form iodine, water, and chloride ions. The overall redox reaction is written by combining the two half equations, first checking that the numbers of electrons balance. Add the two equations together, cancel the electrons, and add state symbols:

  $$ClO^- + 2H^+ + 2e^- \rightarrow Cl^- + H_2O$$
  $$2I^- \rightarrow I_2 + 2e^-$$

  $ClO^-(aq) + 2H^+(aq) + 2I^-(aq) \rightarrow Cl^-(aq) + H_2O(l) + I_2(aq)$

- Some atoms always have the same oxidation number in a **compound**, e.g.
  Group 1 compounds are always +1, e.g. in NaCl the Na is +1 (and Cl is –1).
  Group 2 compounds are always +2, e.g. in MgO the Mg is +2 (and O is –2).
  Group 3 compounds are usually +3, e.g. in Al_2Cl_6 the Al is +3.
  Fluorine is always –1.
  Oxygen is usually –2 (except when with F or in peroxides), e.g. in MgO the O is –2; in OF_2 the O is +2; and in peroxides, e.g. H_2O_2, the O is –1.
  Chlorine is usually –1 (except when with F or O), e.g. in NaCl the Cl is –1; in ClF_3 the Cl is +3; and in NaClO the Cl is +1.
  Hydrogen is usually +1 (except when it is alone with a less electronegative metal), e.g. in HCl the H is +1; but in NaH the H is –1.
  Some metals have one oxidation number; zinc is +2, silver is +1 (usually). Otherwise you will see the **oxidation state** in Roman numerals after the element, e.g. in iron(II) sulfate Fe has oxidation number +2.

- Remember these **rules** to work out the oxidation number of an **element in a formula**:
  1. The sum of the oxidation numbers in a **compound** always equals zero, e.g. magnesium chloride $MgCl_2$: Mg = +2; Cl = –1; + 2 + (2 × –1) = 0
  2. The sum of the oxidation numbers in an **ion** always equals the charge, e.g. chlorate(I) ClO^-: Cl = +1; O = –2; +1 + (–2) = –1.
  3. The oxidation number of an element in its elemental state always equals zero.

- **Bromine** is extracted from **sea water** by using chlorine to oxidise bromide ions. The concentration of bromine is increased by absorption in aqueous SO_2. Re-oxidation by chlorine produces sufficient bromine vapour for condensation to a liquid (see Fig. 4.4).

  The main source of **iodine** is sodium iodate(V) which is in the mineral Chile saltpetre, sodium nitrate. Some species of seaweed extract iodine from seawater. Their ash contains up to 0.5% of iodine.

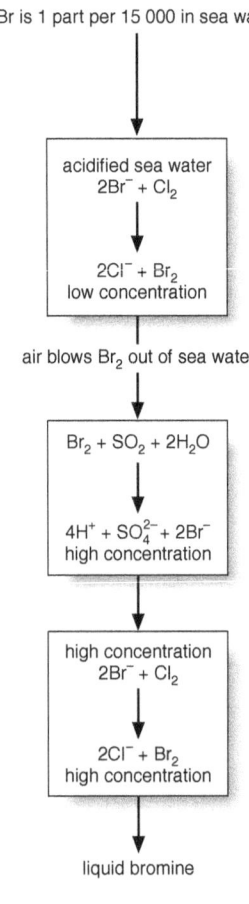

Fig. 4.4 The extraction of bromine from sea water

TESTS

RECALL TEST

1 State the colours of the halogens

   F_2 _____,

   Cl_2 _____,

   Br_2 _____,

   I_2 _____. (2)

2 Why is iodine a solid whereas chlorine is a gas?

   _____
   _____
   _____ (2)

3 Why is the Cl-Cl bond strong relative to the I-I bond?

   _____
   _____ (2)

4 Why is chlorine more reactive than iodine?

   _____
   _____
   _____ (2)

5 Write balanced equations for the reaction of bromine with:

   a iron _____

   b hydrogen _____ (2)

6 Write the names of the compounds made when NaCl is added to concentrated sulfuric acid.

   _____ (2)

7 State the products made when KI is added to concentrated sulfuric acid.

   _____ (3)

8 State the oxidation number of the elements in bold beneath the following formulae:

   MgO **S**O_2 H_2**S**O_3 **S**O_3 H_2**S**O_4 Mg**S**O_4 H_2**S** **N**H_3 **N**H_4^+ **Na**(s) **Cl**$_2$(g)

   _____ (11)

9 Write equations for the following reactions:

   a $Br^-(aq) + Cl_2(aq)$ _____

   b $Cl^-(aq) + I_2(s)$ _____

   c $Ag^+(aq) + Cl^-(aq)$ _____

   d $PCl_5(s) + H_2O(l)$ _____ (4)

   (Total 30 marks)

Unit 4 TESTS

CONCEPT TEST

1 Sodium chlorate(VII) has been used for many years as a weedkiller. It is made in three stages, I, II, and III. I: chlorine is made by electrolysis in a diaphragm cell. II: the chlorine is reacted with warm NaOH when disproportionation occurs. III: $NaClO_3$ is dried by warming to make $NaClO_4(s)$.

   a In the diaphram cell in stage I,

      i identify the electrolyte,

        (1)

      ii write the equation for the reaction at the (+) anode.

        (2)

   b What is meant by the word 'disproportionation'?

      (2)

   c Write the ionic equation for the reaction between warm sodium hydroxide and chlorine. For each chlorine-containing species state the oxidation state of the chlorine.

      (4)

   d The $NaClO_4$ forms according to this equation:

   $4ClO_3 \rightarrow 3ClO_4^- + 2Cl^-$

   Deduce the ionic half equation for the oxidation reaction (assuming alkali is still present).

      (2)

   e What is the test for the chloride ions?

      (2)

2 The amount of iodine in a sample may be determined by titrating with aqueous thiosulfate.

   a Write the equation for the titration reaction.

      (2)

   b Why must starch indicator be used?

      (2)

   c How could the presence of iodide be detected?

      (1)

d When chlorine gas is bubbled through potassium iodide solution iodine-forms.

State two changes that could be observed.

_____

_____ (2)

3 Sulfuric acid reacts with potassium chloride in a different way than with potassium bromide and potassium iodide.

  a Describe the reaction of potassium chloride with sulfuric acid.

    i Write an equation for the reaction,

_____

_____ (1)

    ii What type of reaction is occuring? What would be observed?

_____

_____ (2)

  b Consider the reaction of potassium bromide with sulfuric acid.

    i What would be observed when potassium bromide is added to sulfuric acid? Name the bromine containing product.

_____

_____ (1)

    ii What type of reaction is occuring? Why does potassium bromide react differently to potassium chloride?

_____

_____

_____ (2)

    ii State the oxidation numbers of the sulfur containing reactant and product.

_____

_____ (1)

    iii State the oxidation numbers of the bromine containing reactant and product.

_____

_____ (1)

  c Potssium iodide reacts with sulfuric acid to make different sulfur containing products.

    i What would be observed when potassium iodide is added to sulfuric acid? Note two observations.

_____

_____ (1)

    ii Name the two sulfur containing products.

_____

_____ (1)

(Total 30 marks)

Unit 5 — PERIODIC TABLE

There are different types of atomic orbital (s, p, d, f), but you only have to know the shapes of **s** and **p** orbitals.

- The space occupied by an electron around a nucleus for 95% of the time is called an **atomic orbital**. Up to two electrons can occupy each atomic orbital.

  s orbitals are spherical with the nucleus at the centre (see Fig. 5.1).

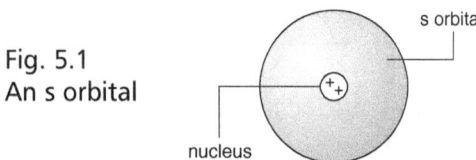

Fig. 5.1 An s orbital

p-orbitals are shaped like two balls stuck together, with the nucleus where the spheres touch (see Fig. 5.2).

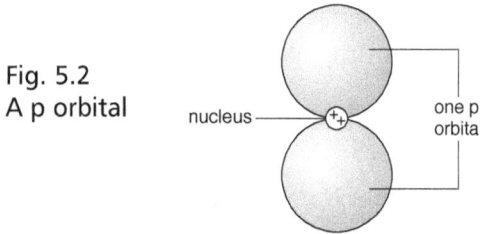

Fig. 5.2 A p orbital

Orbitals of the same type group together in **subshells**. There is one s orbital in a s-subshell, three p orbitals in a p-subshell, and five d orbitals in a d-subshell.

- **Shells** contain groups of subshells that have similar energies. In a given shell, the energies of the subshells increase in the order s < p < d (see Fig. 5.3).

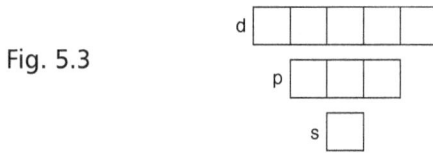

Fig. 5.3

- A list of an atom's occupied subshells is called its **electronic configuration**. You may be asked to write the configuration for any of the elements 1 (H) to 36 (Kr), e.g. Kr = $1s^2\ 2s^2\ 2p^6\ 3s^2\ 3p^6\ 3d^{10}\ 4s^2\ 4p^6$. The term $3p^6$, for example, means that there are 6 electrons in the p subshell that is part of the shell n-=-3. Note that 4s is filled before 3d (see unit 24).

- Use the periodic table to help you write electronic configurations. The group 1 and 2 metals are in the **s-block** because the s-orbital is being filled. Similarly, the block on the right (groups 3–0) is called the **p block**. The period (row) number tells you the principle quantum number of the s or p-subshell being filled.

- You must recall that the electronic configuration determines the **chemical properties** of an element.

- **Covalent bonds** form when atomic orbitals overlap to make a **bonding molecular orbital**. Electrons in the molecular orbitals are shared between the atoms.

  A single bond is made when two orbitals overlap end on. This bond is called a **sigma (σ) bond** (see Fig. 5.4).

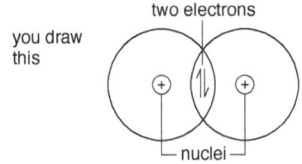

Fig. 5.4 A σ bond

  Two p orbitals overlap sideways to make a single **pi (π) bond** (see Fig. 5.5).

  A **double bond** consists of two bonds – a sigma and a pi bond. The sigma bond is symmetrical around the line joining the two nuclei. The pi bond exists either side of the sigma bond.

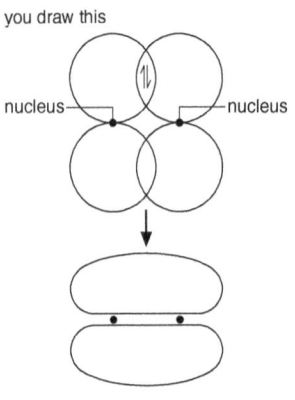

Fig. 5.5 A π bond

- You must recall the definition:
  1st ionisation energy: $X(g) \rightarrow X^+(g) + e^-$ The energy required to remove one mole of electrons from one mole of gaseous atoms to form one mole of gaseous ions with a single positive charge.
  You may have to define successive ionisation energies of an element, for example:
  3rd ionisation energy: $X^{2+}(g) \rightarrow X^{3+}(g) + e^-$ The energy required to remove one mole of electrons from one mole of gaseous X^{2+} ions to form one mole of gaseous X^{3+} ions.
  It takes energy to pull the (−) e from the (+) ion, so all these changes are **endothermic**. As successive electrons are removed the cation charge increases so the ionisation energies increase.

- Examine the graph of the 1st ionisation energies of **successive elements** (Fig. 5.7).

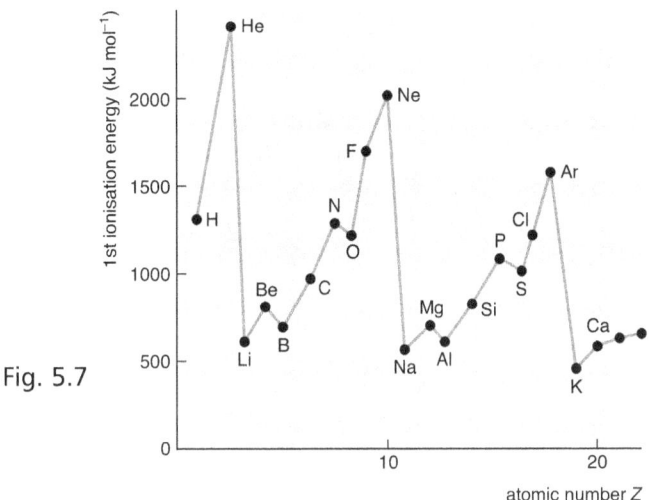

Fig. 5.7

Note:
(1) the general **increase** in 1st ionisation energy **from Na to Ar**, due to increased proton number. In a similar way there is a general increase in 1st ionisation energy **from Li to Ne.**

(2) the **peaks** in 1st ionisation energy are always the **noble gases** because these have the highest proton number for the period, just before a new shell starts in the next period.

(3) there is a **drop** in 1st ionisation energy between **group 2 and 3** elements. The electron in the group 3 atom is lost more easily from the p-orbital, which is further from the nucleus, than the electron lost from the s-orbital in the group 2 atom.

(4) there is a **drop** between **groups 5 and 6**. Group 6 atoms have two electrons paired in the p subshell. Repulsion between these electrons makes one of them easier to remove.

Unit 5 — PERIODIC TABLE

- The elements are arranged in the periodic table in order of **increasing proton number** (atomic number). The periods (rows) show **repeating** physical and chemical properties (this is known as **periodicity**). The elements in a given group (column) have similar properties because the outer shells of electrons have similar structures.

- The **atomic radius** decreases across a period as the proton number increases and pulls in the outer electrons (see Fig. 5.8).

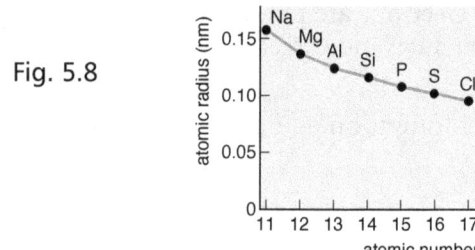

Fig. 5.8

For the same reasons, the **cation radius** decreases as the nuclear charge increases, while the electronic configurations of the common ions in a period are similar (see Fig. 5.9).

The **anions** in a period have one more complete shell than the cations, making the anions larger, while the **anionic radius** decreases as nuclear charge increases.

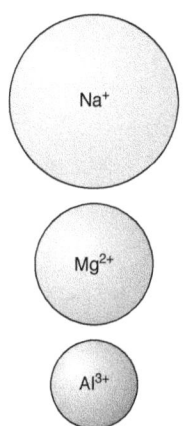

Fig. 5.9 The relative sizes of the Na^+, Mg^{2+}, and Al^{3+} ions

- The **electrical conductivity** shows a periodic pattern which is explained by the ideas of bonding and structure studied in unit 1.

  The elements on the left (groups 1 to 3) are very good conductors because they are metals (see unit 1).

  In group 4 the element carbon is a moderately good conductor when in the form of graphite, but a poor electrical conductor when in the form of diamond. Also in group 4, silicon is a moderately good electrical conductor, though here conductivity is due to impurities. Graphite, diamond, and silicon are giant covalent structures.

  The elements on the right (in groups 5 to 0) are poor conductors due to their simple covalent structure and bonding, so there are no free electrons to carry the charge.

- The proton number increases **across a period** (row) and so the nucleus becomes more attractive to electrons. This trend explains the increases in electronegativity, 1st ionisation energy, and 1st electron affinity.

  The proton number increases **down a group** and so the number of filled electron shells increases, which (i) increases the distance between the nucleus and the outer electrons and (ii) increases shielding of the nuclear charge. The atomic radius increases while the nuclear attraction for the outer electrons decreases, causing electronegativity and 1st ionisation energy to decrease.

TESTS
RECALL TEST

1. What is meant by an 'atomic orbital'?

   _____

   _____ (2)

2. Draw:

   a a s orbital (1)

   b a p orbital (1)

   c a sigma bond (1)

   d a pi bond (1)

3. Write the electronic configuration of krypton.

   _____ (2)

4. What is meant by 's-block element'?

   _____

   _____ (1)

5. State what determines the chemical properties of an element.

   _____ (1)

6. Why does electronegativity increase across the third row of the periodic-table?

   _____ (1)

7. State and explain the change in 1st ionisation energy across the third-row-of-the periodic table.

   _____

   _____

   _____ (1)

8. In group 2, with increasing atomic number, state whether the follow increase, decrease, or stay the same:

   a the atomic radii _____ (1)

   b electronegativity _____ (1)

   c 1st ionisation energy _____ (1)

9. Define '1st ionisation energy'.

   _____

   _____

   _____ (1)

10. Why is the 2nd ionisation energy always more endothermic than the 1st ionisation energy for a particular element?

    _____

    _____

    _____ (2)

TESTS

11 Define the 4th ionisation energy.

_____

_____

_____ (2)

(Total 30 marks)

12 Which of the first 20 elements (H to Ca) has:

   a the highest 1st ionisation energy _____ (1)

   b the weakest Van der Waals forces _____ (1)

   c the smallest cationic radius _____ (1)

   d the smallest anionic radius _____ (1)

13 For the third period explain why the melting point of the elements starts high (for Na, Mg, Al) then peaks at Si, but is lower for P, S, Cl, Ar.

_____

_____

_____

_____

_____

_____

_____ (6)

CONCEPT TEST

1 This question concentrates on ionisation energy.

   a Give an equation for the first ionisation energy of chlorine atoms.

_____

_____ (2)

   b The first ionisation energy of helium is the highest of all atoms. Explain why this is so.

_____

_____ (2)

   c Give equations that represent the first ionisation energy and second ionisation energy of sulfur.

   1st ionisation energy of S:

_____

   2nd ionisation energy of S:

_____ (2)

   d In terms of ionisation energy, state and explain the trend in the reactivity of the group 2 elements.

_____

_____

_____ (3)

2 State which element has the higher ionisation energy (IE) for each pair of elements. Give a short explanation in each case.

   a 1st IE of C and 1st IE of Si. _____

   _____

   b 1st IE of Ar and 1st IE of K. _____

   _____

   c 1st IE of Be and 1st IE of B. _____

   _____

   d 1st IE of Na and 1st IE of Mg. _____

   _____

   e 2nd IE of Na and 1st IE of Mg. _____

   _____(10)

3 a Draw a graph of the log of successive ionisation energies of potassium.

(4)

   b Explain why the successive ionisation energies of potassium generally increase.

   _____

   _____(1)

   c Explain the shape of the graph.

   _____

   _____

   _____

   _____(4)

   d Why are there only 19 ionisation energies for potassium?

   _____(2)

(Total 30 marks)

Unit 6: ORGANIC BONDING AND ISOMERS

- Organic compounds are based on **skeletons** of **carbon atoms** covalently bonded together in **chains** and **rings**. (See unit 1.) When different organic molecules have similar structures and react similarly, then we list them in a group called a **homologous series**. The members of each series have similar structures but different numbers of $-CH_2-$ groups.

> **Alkanes** have the general formula C_nH_{2n+2}. They are called **saturated** hydrocarbons because all the C-C bonds are single bonds.

- The members of the **alkane** homologous series consist of only carbon and hydrogen atoms joined by single covalent bonds. You need to recall the first ten alkane names and formulae. (The general formula C_nH_{2n+2} will help).

Number of C atoms	Name	Formula
1	methane	CH_4
2	ethane	C_2H_6
3	propane	C_3H_8
4	butane	C_4H_{10}
5	pentane	C_5H_{12}
6	hexane	C_6H_{14}
7	heptane	C_7H_{16}
8	octane	C_8H_{18}
9	nonane	C_9H_{20}
10	decane	$C_{10}H_{22}$

The **boiling points** and the **melting points** of the alkanes increase with increasing number of carbon atoms.

Straight chain molecules have higher boiling points than **branched isomers** of the same size. **Example:** pentane C_5H_{12} b.p.= 36 °C; methylbutane C_5H_{12} b.p.= 28 °C.

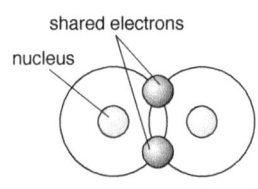

The nucleus is very close to the shared electrons so the opposite charges attract strongly.

Fig. 6.1

- The **alkanes** are unreactive because they contain strong C-C and C-H bonds. The **bonds** are **strong** (have a high average bond enthalpy – see unit 12) because the atoms are very small and the outermost electrons are not shielded from the attraction of the nuclear charge. This effect causes the C-C bond, for example, to be **short** and strong (see Fig. 6.1).

- The **alkenes** contain C=C double bonds. They are a more reactive series of compounds than the alkanes. A double bond consists of a strong sigma bond and a weaker pi bond (see unit 1). Electrons in the sigma bond are concentrated between the two nuclei; electrons in the pi bond are concentrated further away and to the sides of the nuclei, resulting in a weaker bond (see Fig. 6.2).

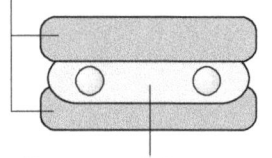

The electrons in the pi bond are far from the nuclei.

The electrons in the sigma bond are close to the nuclei.

Fig. 6.2

- The **halogenoalkanes** are a homologous series of compounds that consist of alkanes which have one or more hydrogen atoms replaced by a halogen atom. Halogen atoms are approximately 2 to 4 times larger than hydrogen atoms. The nucleus of the large halogen atom is far from the shared electrons in the carbon–halogen bond. For example, the bromine atom is so large that the C-Br bond is just two-thirds the strength of the C-H bond (see Fig. 6.3).

Fig. 6.3

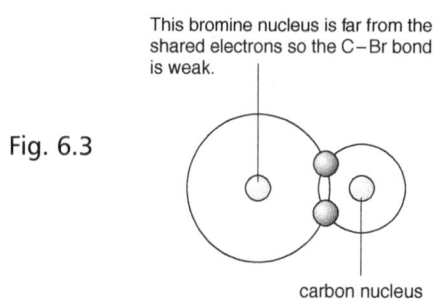

This bromine nucleus is far from the shared electrons so the C–Br bond is weak.

carbon nucleus

- **Alkanes** are so **unreactive** that you must use harsh conditions to make them react. Ultraviolet light has high energy that will break bonds in alkanes and start a reaction (see unit 8 for details). High temperature can also be used to initiate reactions, as in the combustion of petrol (consisting mostly of alkanes).

  Alkenes are more **reactive** than alkanes. Their reactions often take place at 'room temperature'.

  Organic reactions are usually **slow**. Heating a flask of chemical reagents allows volatile liquids to boil and escape. Stoppering the heated flask would cause an explosion. The solution is to fit a **vertical condenser** to a heated flask. This arrangement allows extended boiling without loss of volatile substances. In exams you need to write this condition as **heat under reflux** (See Fig.6.4).

  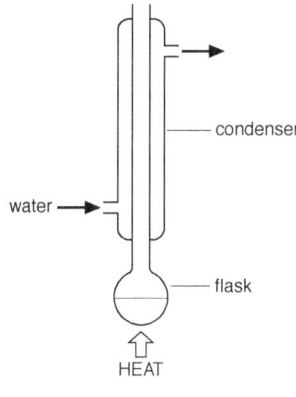

  Fig. 6.4 Heat under reflux

  Some **halogenoalkanes** (chloro-, bromo-, and iodoalkanes) have weaker covalent bonds than alkanes so **react faster**. They also react faster because the bonds are easily polarised. However, they still need a little help, so the reaction mixture is usually heated under reflux.

- The presence of C=C and C-Br bonds result in a reactive site on a molecule, so the C=C and C-Br are called **functional groups**. You do need to learn these functional groups as soon as possible. You could start with the top six in the table below. The functional groups are shown in bold type.

Homologous series	Name of example	Graphical formula	Linear abbreviated formula
alkanes	propane		$CH_3CH_2CH_3$
alkenes	propene		$H_2\mathbf{C=C}HCH_3$
halogenoalkanes	1-bromopropane		$CH_3CH_2CH_2\mathbf{Br}$
alcohols	propan-1-ol		$CH_3CH_2CH_2\mathbf{OH}$
aldehydes	propanal		$CH_3CH_2\mathbf{CHO}$
ketones	propanone		$CH_3\mathbf{CO}CH_3$
amines	1-aminopropane		$CH_3CH_2CH_2\mathbf{NH_2}$
nitriles	propanenitrile		$CH_3CH_2\mathbf{CN}$
carboxylic acids	propanoic acid		$CH_3CH_2\mathbf{COOH}$
carboxylic acid salts	sodium propanoate		$CH_3CH_2\mathbf{CO_2^-Na^+}$
esters	ethyl propanoate		$CH_3CH_2\mathbf{COO}CH_2CH_3$
amides	propanamide		$CH_3CH_2\mathbf{CONH_2}$
carbonyl chlorides	propanoyl chloride		$CH_3CH_2\mathbf{COCl}$

Unit 6 — ORGANIC BONDING AND ISOMERS

- Look at the structures of 1-bromopropane and 2-bromopropane (see Fig. 6.5). Both have the same numbers of atoms of C, H, and Br, and the same formula C_3H_7Br. These compounds have different structures, so we call them **structural isomers**.

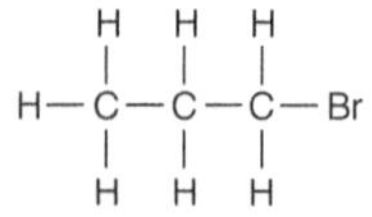

Fig. 6.5

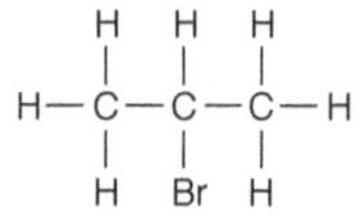

Some compounds containing **C=C double bonds** exhibit **E-Z isomerism**. Look at the structures of but-2-ene (see Fig. 6.6). The atoms are joined together in the same order $CH_3CHCHCH_3$ but the molecules have different structures because the two methyl $-CH_3$ groups are either on the same side or on different sides of the C=C double bond.

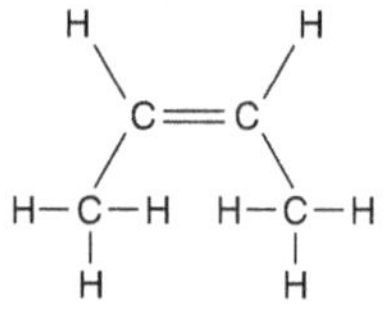

Fig. 6.6

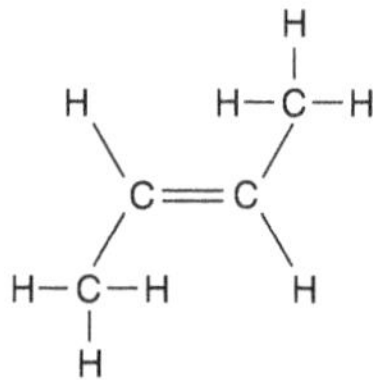

Alkanes do not exhibit geometrical isomerism because the atoms can **rotate** around a single C-C bond. There are no geometrical isomers of butane $CH_3CH_2CH_2CH_3$. **Alkenes** show this type of isomerism because of the **restricted rotation** around the C=C bond.

- You should know how to write or draw **molecular**, **structural**, **linear**, **display** or **graphical**, and **skeletal formulae** (see Fig. 6.7).

- Note that you may be asked to draw cyclic compounds. In Fig. 6.7 there is the example of cyclohexane which has the formula C_6H_{12}.

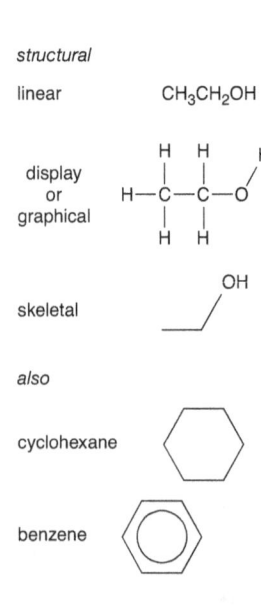

Fig. 6.7 Examples of different formula types

TESTS
RECALL TEST

1 What is a homologous series? _____ (1)

2 Name and write the formulae of the first ten alkanes.

_____

_____

_____

_____

_____ (3)

3 Why are the alkanes so unreactive?

_____

_____ (1)

4 Why are alkenes very reactive?

_____ (1)

5 Why are halogenoalkanes more reactive than alkanes?

_____

_____ (1)

6 What sort of conditions are required for the following to react?

   a alkanes _____

   b alkenes _____

   c halogenoalkanes _____ (3)

7 Explain

   a structural isomerism,

   _____

   _____ (1)

   b E-Z isomerism.

   _____

   _____ (1)

8 How many structural isomers may be made from C_4H_8? _____ (1)

9 How many isomers may be made from C_4H_7Br? _____ (1)

10 In a covalent molecule, how many bonds do these make?

   a carbon ___   b nitrogen ___   c oxygen ___   d fluorine ___   (4)

Unit 6 — TESTS

11 State and explain the change in the strength of these covalent bonds: C-F, C-Cl, C-Br, C-I.

_____

_____

_____ (2)

12 For each of the functional groups in the table on page 27, state the intermolecular force (or bonding) present that determines the boiling point of the substance.

_____

_____

_____

_____

_____

_____

_____ (7)

13 Show the graphical (see Fig. 6.7) and linear formula of propan-2-ol and then show how it can be simplified to the skeletal formula.

(3)
(Total 30 marks)

CONCEPT TEST

1 When a solution of bromine is shaken with hexene, C_6H_{12}, the bromine is decolorised. However, when bromine is added to hexane in the dark there is no decolorisation.

a Write an equation for the reaction of hexene with bromine.

_____ (1)

b Explain, in terms of the bonding, why no reaction occurs when a solution of bromine is shaken with hexane in the dark.

_____

_____

_____ (2)

2 Suggest the conditions required for these three reactions:

a Ethane may be mixed with HBr to form bromoethane.
_____

b Bromoethane will react with NaOH(aq) to form ethanol.
_____

c Bromoethane may also be formed from ethane.
_____ (3)

3 (See also unit 11) Consider these bond enthalpies.

a Explain the trend in the halogen–hydrogen bond enthalpies.
_____
_____
_____
_____ (3)

Bond	Enthalpy (kJ mol^{-1})
F-H	562
Cl-H	431
Br-H	366
I-H	299
C-C	348
C=C	612
Si-Si	176

b Explain the difference between the C-C and C=C bond enthalpies.
_____
_____
_____
_____
_____ (3)

c Explain why the Si-Si bond is weaker than the C-C bond.
_____
_____
_____ (3)

4 The compound right is being considered as an insecticide:

a Draw the *E-Z* isomers that this compound may have.

it may be drawn like this:

b Explain what makes this molecule so reactive.
_____
_____
_____
_____
_____ (5)

(Total 20 marks)

Unit 7: ORGANIC MECHANISMS

You will find studying organic chemistry easier if you know which type or class a reaction belongs to. Recognising the **reaction class** helps you choose the correct reagents.

- There are millions of possible organic reactions but most of them fall into one of the ten classes described here. Each type of reaction proceeds from reactants to products in a distinct series of steps, known as the **reaction mechanism**.

- **Acid–base reactions** happen when an organic molecule **donates** a proton H^+ or accepts a proton. (See Fig. 7.1.)

  When a molecule **donates** a proton, then it is acting as an **acid**. An **acidic hydrogen atom** leaves the molecule as H^+ and is replaced by a metal ion (or NH_4^+). Example: $CH_3COO\mathbf{H} + Na\mathbf{OH} \rightarrow CH_3COO^-Na^+ + \mathbf{H_2O}$

  Example: $CH_3COOH + NaHCO_3 \rightarrow CH_3COO^-Na^+ + H_2O + CO_2$

  CH_3COOH (ethanoic acid) is acting as an acid because it donates H^+ to the base OH^-. Sodium hydrogencarbonate and carbonates can also act as bases.

  When a molecule **accepts** a proton H^+, then it is acting as a **base**.
  Example: $CH_3\mathbf{NH_2} + \mathbf{H}Cl \rightarrow CH_3\mathbf{NH_3^+}Cl^-$

  CH_3NH_2 (methylamine) is acting as a base because the amino $-NH_2$ group is accepting a proton H^+ from the aqueous hydrochloric acid HCl.

Acid: donates H^+
Base: accepts H^+

Fig. 7.1

Fig. 7.2

- A molecule is **oxidised** when it **gains O** or **loses H**.
  Example: If you leave an unfinished bottle of wine opened overnight, then atmospheric oxygen oxidises the alcohol CH_3CH_2OH (ethanol) to vinegar CH_3COOH (ethanoic acid). The alcohol molecule has lost two H atoms and gained one O. In the laboratory, alcohols are oxidised by heating under reflux with an **oxidising agent**.

  A molecule is **reduced** when it **loses O** or **gains H**.
  Example: Ethanal CH_3CHO gains two H atoms when it is reduced to ethanol CH_3CH_2OH. A powerful **reducing agent** is needed. (See Fig. 7.2.)

- **Condensation reactions** occur when two or more molecules join and a small molecule is given off (often H_2O or HCl). Condensation is **addition** followed by **elimination**. Example: Ethanoic acid CH_3COOH and ethanol CH_3CH_2OH condense to form the ester $CH_3COOCH_2CH_3$ with the elimination of water H_2O, which is known as esterification. (See Fig. 7.3.)

Condensation: two molecules join and small molecule given off
Hydrolysis: break bond + H_2O

Fig. 7.3

- **Hydrolysis reactions** occur when **water** breaks a bond in a molecule.
  Example: Heating the **ester** ethyl ethanoate with water splits it into **ethanoic acid** and **ethanol**. An equilibrium mixture forms very slowly.
  $CH_3COOCH_2CH_3 + H_2O \rightleftharpoons CH_3COOH + CH_3CH_2OH$

- **Hydration reactions** involve the **addition** of **water** to a molecule.
  Example: $H_2O + CH_2CH_2 \rightarrow CH_3CH_2OH$ (See Fig. 7.4.)

Dehydration $- H_2O$
Hydration $+ H_2O$

Fig. 7.4

- **Dehydration reactions** occur when **water** is **removed** from a molecule.
  Example: **Ethanol** is dehydrated to **ethene** when heated with the **dehydrating agent**. Each molecule loses 2 H atoms and one O atom.
  $CH_3CH_2OH \rightarrow CH_2CH_2 + H_2O$

- **Substitution reactions** occur when one group of atoms is replaced by another group. Example: $CH_3CH_2Br + OH^- \rightarrow CH_3CH_2OH + Br^-$

  Elimination reactions occur when some **atoms** are **removed** from an organic molecule. Remember that the elimination of **water** is called **dehydration**. You need to know about only one type of elimination reaction. Example: Heating **bromoethane** with KOH in pure ethanol forms **ethene** by the elimination of HBr.

  $CH_3CH_2Br + KOH \rightarrow CH_2CH_2 + H_2O + KBr$

- **Addition reactions** occur when a group of atoms are added to a molecule and no atoms are lost. **Example: Hydrogen bromide** adds to **ethene** to make **bromoethane**. $CH_2CH_2 + HBr \rightarrow CH_3CH_2Br$

- A **reaction mechanism** explains how a reaction happens. It shows how **electrons move** and which **bonds form** or **break** in each of the **steps** that make up the overall reaction. When you know the mechanism you generally realise which **reagents** are needed. Many reagents can be classed as nucleophiles or as electrophiles.

- A **nucleophile** is a molecule or ion with an **electron-rich** site which can **donate** a pair of **electrons**. Nucleophiles **attack** the δ+ C atom of **weak bonds** in halogenoalkanes, aldehydes, or ketones. (See Fig. 7.5 for the nucleophilic substitution of bromomethane).

Nucleophiles include CN^-, OH^-, and Cl^-, as well as H_2O and NH_3. All nucleophiles have **lone pairs** which they **donate** to form **new bonds**.

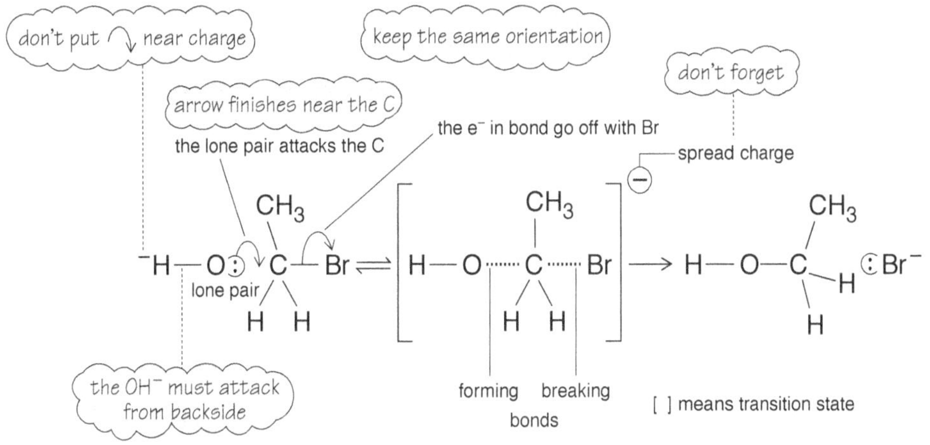

Fig. 7.5

In the substitution reaction given above, the **OH⁻** ion is acting as a **nucleophile**. The reaction is therefore classed as a **nucleophilic substitution** reaction.

- An **electrophile** is a molecule or ion with an **electron-deficient** site which can **accept** a pair of **electrons**. Electrophiles react with the electron-rich pi bonds found in alkenes and aromatic molecules. (See Fig. 7.6 for the electrophilic addition of HBr to ethene.)

Electrophiles include positive ions such as H^+ and NO_2^+ and molecules such as Br_2 and HBr.

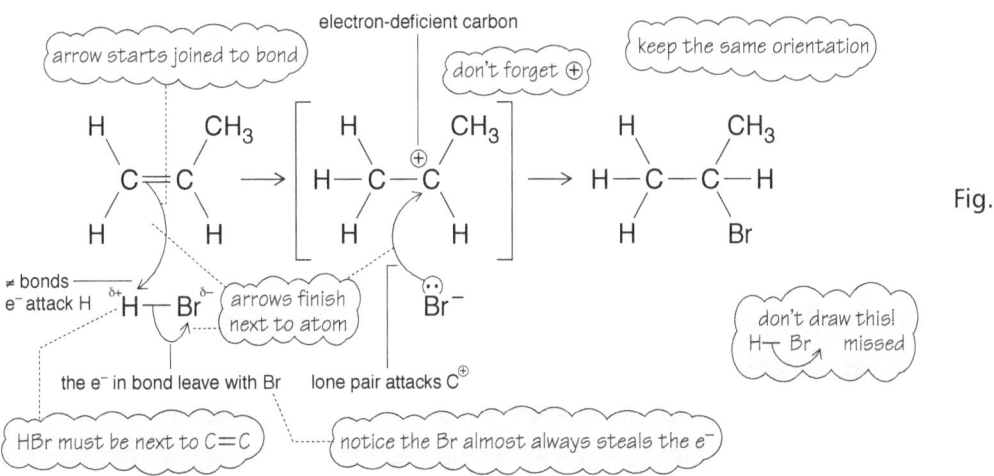

Fig. 7.6

In the addition reaction given above, the **HBr** is acting as an **electrophile**. The reaction is therefore classed as an **electrophilic addition** reaction.

Unit 7 — ORGANIC MECHANISMS

- You will notice that when **bonds break**, the reaction mechanism usually shows that the **bonding pair** of electrons become located on **one atom**. This type of bond breaking is called **heterolytic fission**. Nucleophilic and electrophilic reactions usually involve heterolytic fission (see Fig. 7.7).

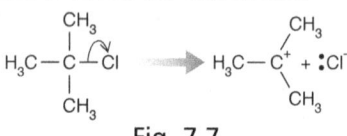

Fig. 7.7

- To show the **movement** of **two electrons** in a mechanism, you draw a **double-barbed curly arrow** (⌒). Ensure you position the tail of the arrow on the lone pair, covalent bond, or pi bond and position the point exactly where the pair of electrons ends up.

- A **free radical** is a molecule or atom with an **unpaired electron**.

- Free radical reactions involve the **movement** of **single electrons**. To show the movement of one electron, you draw a **single-barbed curly arrow** called a fish hook (⌒). When a bond breaks in a free radical reaction, the bonding electrons separate and one electron goes to each atom. This type of bond breaking is called **homolytic fission** (see Fig. 7.8).

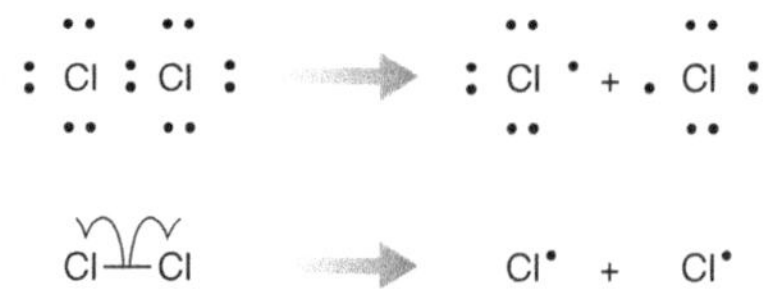

Fig. 7.8

- A free radical is shown as a dot against the symbol or formula e.g. Br• or $CH_3^\bullet$.

- Alkanes react with halogens in ultraviolet light to make **halogenoalkanes**. An example is the bromination of methane.

  $CH_4 + Br_2 \rightarrow CH_3Br + HBr$

  Further substitution happens, producing CH_2Br_2, etc. The mechanism is by **free radical substitution**, which takes place in three steps: 1 **initiation**; 2-**propagation**; 3 **termination**.

  Initiation produces free radicals by the heterolytic fission of the Br-Br bond.
  $Br_2 \rightarrow 2Br^\bullet$

  Propagation occurs when free radicals react with molecules to make other free radicals. Here, the Br• radical strikes the H atoms on the outside of the molecule (not the C atom):
  $CH_4 + Br^\bullet \rightarrow CH_3^\bullet + HBr$
  The product then forms, together with further free radicals.
  $CH_3^\bullet + Br_2 \rightarrow CH_3Br + Br^\bullet$
  Propagation is called a **chain reaction** because it uses up and then produces the reactive Br• radical.

  Termination happens when free radicals combine to produce neutral molecules, e.g.
  $2CH_3^\bullet \rightarrow CH_3CH_3$
  $CH_3^\bullet + Br^\bullet \rightarrow CH_3Br$

TESTS
RECALL TEST

1. Name the type of reagent required to carry out these changes:
   a. turn CH_3COOH into $CH_3COO^-Na^+$ _____
   b. turn CH_3NH_2 into $CH_3NH_3^+$ _____
   c. turn $C_6H_5CH_3$ into C_6H_5COOH _____
   d. turn CH_3CN into $CH_3CH_2NH_2$ _____
   e. turn CH_3CN into CH_3CONH_2 _____
   f. turn CH_3CONH_2 into CH_3CN _____ (6)

2. State the type of reaction:
   a. $CH_3CH_2OH + CH_3COCl \rightarrow CH_3COOCH_2CH_3 + HCl$
   _____ (1)
   b. $CH_3CONHCH_2CH_3 + NaOH \rightarrow CH_3CO^-Na^+ + CH_3CH_2NH_2$
   _____ (1)

3. What do these words mean?
   a. nucleophile
   _____
   _____ (2)
   b. electrophile
   _____
   _____ (2)
   c. addition
   _____
   _____ (2)
   d. substitution
   _____
   _____ (2)

4. What type of mechanism occurs in each of these reactions?
   a. $CH_3CH_2Cl + NaOH \rightarrow CH_3CH_2OH + NaCl$
   _____ (2)
   b. $CH_3CHCH_2 + HBr \rightarrow CH_3CHBrCH_3$
   _____ (2)

   (Total 20 marks)

Unit 7 TESTS

CONCEPT TEST

1 a Give an example of each of these reagent types:

Oxidising agent _____

Reducing agent _____

Dehydrating agent _____ (3)

b Explain why converting ethanol into ethene is called dehydration.

_____

_____ (1)

c Which reagent will convert ethanoic acid into sodium ethanoate?

_____

_____ (1)

d How may sodium ethanoate be converted back to ethanoic acid?

_____

_____ (1)

2 Ethanoic acid and ethanol react slowly to make ethyl ethanoate.

a What type of reaction does this illustrate?

_____

_____ (1)

b Ethyl ethanoate may be split back into its components. Which chemical must be present for this to happen?

_____

_____ (1)

3 a Explain what is meant by 'nucleophilic substitution'.

_____

_____ (2)

b Explain what is meant by 'heterolytic fission'.

_____

_____ (2)

c Draw the mechanism of the reaction between bromoethane and potassium cyanide, KCN, which reacts in a similar way to sodium hydroxide, NaOH.

(3)

4 Bromine, Br₂, reacts with ethene (containing >C=C<), but does not react with propanone (containing >C=O), which has a similar structure.

   a Draw the mechanism for the reaction between ethene and bromine.

   (3)

   b Why does >C=C< react with electrophiles and >C=O react with nucleophiles?

   _____
   _____
   _____
   (2)
   (Total 20 marks)

Unit 8 — ALKANES AND ALKENES

> The sources of alkanes are natural gas and crude oil (petroleum). **Natural gas** consists mostly of **methane**. **Crude oil** is a mixture of hundreds of different hydrocarbons, most of which are **alkanes**.

> R, R', and R" all indicate an alkyl chain, e.g. a methyl group $-CH_3$.

> LRF (lead replacement fuel) has replaced leaded petrol.
> LPG (liquid petroleum gas) contains butane.
> Natural gas contains methane.
> Gasohol is an ethanol–petrol mix.

- The alkanes are very **unreactive** (see unit 6). High-energy conditions are required to initiate reactions, that is **high temperature** (e.g. for combustion) or **ultraviolet light** (in free radical substitution reactions which was discussed at the end of unit 7).

  Alkanes **combust** (burn) in air or oxygen. **Complete combustion** takes place in **excess oxygen** to form carbon dioxide and water. For example, in the case of methane:

  $CH_4 + 2O_2 \rightarrow CO_2 + 2H_2O$

  Incomplete combustion takes place in **limited oxygen** to form poisonous carbon monoxide and water. For example, in the case of methane:

  $2CH_4 + 3O_2 \rightarrow 2CO + 4H_2O$

- Combustion of fossil fuels causes atmospheric pollution.

  The carbon dioxide formed is a "greenhouse gas" in that it causes atmospheric global warming. Other greenhouse gases are methane and water vapour. These gases absorb infra-red radiation that would otherwise escape into space, so causing the atmosphere to heat up.

  Most fossil fuels contain sulfur compounds as impurities. Combustion converts the sulfur into sulfur dioxide which contributes to acid rain, and causes lung disease.

- Petrol is a fraction from the crude oil distillation and made by catalytic cracking, but unimproved petrol would damage car engines by combusting too early, which is called **pre-ignition**. Originally, lead compounds were added to limit pre-ignition, but lead harms the nervous system, and poisons catalytic converters.

  Catalytic converters in car exhaust systems use **Rh** on a ceramic honeycomb (to **economize** on expensive Rh) to catalyse the conversion of pollutants (**CO**, **NO$_x$**, unburnt **hydrocarbons**) into **CO$_2$**, **N$_2$**, and **H$_2$O**.

  Branched alkanes, cycloalkanes, and arenes are used in petrol to encourage efficient combustion. These chemicals are made by catalytic cracking, catalytic reforming, and isomerisation. Also methanol is added to improve combustion.

- As crude oil becomes scarce and expensive, it will be replaced by new fuels. Countries without sources of crude oil use alcohol, produced from plants, in their 'biofuels'. Ethanol or methanol could replace petrol, though ethanol combustion produces less energy per kilogram. Also if ethanol is made from grain then world food prices would increase.

  Cheaper to run than petrol, gaseous fuels are already used, but do require bulky high-pressure fuel tanks. Hydrogen may be the fuel of the future because it is non-polluting, but it requires heavy fuel tanks. Solar, wind, tidal, and nuclear energy could supply the energy required to manufacture hydrogen.

- **Fractional distillation** separates crude oil into groups of hydrocarbons called **fractions**. The temperature **decreases** up the fractionating column. Fractions containing low b.p. gases come from the top of the column; high b.p. substances that solidify at room temperature come from the bottom (see Fig. 8.1).

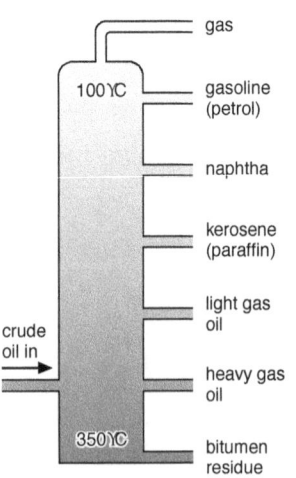

Fig. 8.1 The main fractions from the distillation of crude oil

Excessive amounts of the solid bitumen residue and other long-chain hydrocarbons are usually produced. **Cracking** breaks them into more useful molecules with shorter chains and also produces ethene.

Thermal cracking uses high temperature and pressure to split long-chain alkanes into short-chain alkanes and a high proportion of alkenes (by a free radical mechanism). Hydrogen is a useful by-product.

Catalytic cracking uses a slight pressure, high temperature, and zeolite catalysts to split long-chain alkanes into fractions used to make petrol, together with arenes ('aromatic' hydrocarbons containing benzene rings) (by a carbon cation (C^+) mechanism).

Catalytic reforming is a similar process to catalytic cracking. The process also produces **cyclic** and **aromatic** compounds.

Isomerisation is one form of reforming. It converts straight-chain alkanes into **branched-chain alkanes** which improves the octane number of petrol to stop its pre-ignition, which damages engines.

- **Alkenes** react by **electrophilic addition**, which is explained in unit 7.
  Alkenes have the general formula C_nH_{2n}. They are very **reactive** because they contain **unsaturated** C=C double bonds (C=O and C≡N etc. bonds are also unsaturated). Reactions usually happen at room temperature.

- You must recall these reactions of ethene. They all occur by electrophilic addition.
  $CH_2CH_2 + HBr \rightarrow CH_3CH_2Br$ (room temperature)
  $CH_2CH_2 + Br_2 \rightarrow CH_2BrCH_2Br$ (room temperature)
  $CH_2CH_2 + Br_2(aq) \rightarrow CH_2BrCH_2OH + HBr$ (room temperature)
  This last reaction acts as a **test** for alkenes, because the red-brown colour of Br_2 rapidly disappears.
  Also,
  $CH_2CH_2 + H_2O \rightarrow CH_3CH_2OH$ using an acid catalyst H_3PO_4 and steam.

- The following is a useful explanation which is usually examined at A2.
  During **electrophilic addition** the intermediate forms a C^+ ion, called a **carbocation**. If the C^+ is **tertiary** (e.g. $(CH_3)_3C^+$, which has three -CH_3 groups attached to it) then the C^+ ion is relatively stable and the reaction proceeds steadily. The -CH_3 groups donate electron density which spreads out and stabilises the + charge. The **secondary** C^+ ion (e.g. $(CH_3)_2HC^+$) is less stable, and the **primary** C^+ ion (e.g. $CH_3H_2C^+$) is the least stable.

- **Hydrogenation** uses a **nickel catalyst** at 200°C and high pressure to add hydrogen atoms across the alkene double bond.
  $CH_2CH_2 + H_2 \rightarrow CH_3CH_3$
  Hydrogenation changes liquid vegetable oils into solid **margarine**.

Unit 8 — ALKANES AND ALKENES

- Alkenes are used to make **addition polymers**, e.g. polyethene is made from ethene using a Ziegler-Natta catalyst, high temperature, and high pressure. $n\text{CH}_2\text{CH}_2 \rightarrow [\text{-CH}_2\text{CH}_2\text{-}]_n$ (n is a large number 2000–35 000).

- Other monomers (substituted alkenes) make a wide range of addition polymers. (See Fig. 8.4.)

Fig. 8.4

Monomer name	Monomer structure	Polymer repeating unit	Polymer name	Polymer uses
ethene	CH_2CH_2	$[\text{-CH}_2\text{CH}_2\text{-}]_n$	poly(ethene)	plastic bags
chloroethene (vinyl chloride)	CH_2CHCl	$[\text{-CH}_2\text{CHCl-}]_n$	poly(chloro-ethene) (PVC)	flooring, clothes, pipes
propene	CH_2CHCH_3	$[\text{-CH}_2\text{CH(CH}_3\text{)-}]_n$	poly(propene)	plastic bottles
tetrafluoroethene	CF_2CF_2	$[\text{-CF}_2\text{CF}_2\text{-}]_n$	poly(tetrafluoro-ethene) (PTFE)	non-stick coating

- Be careful when drawing adition polymers. Only two carbons are in the chain of the repeating unit:

  draw this:
  $$\left[\begin{array}{cc} H & CH_3 \\ | & | \\ -C-C- \\ | & | \\ H & H \end{array} \right] \checkmark$$

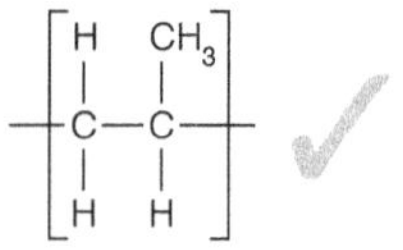

  and NOT this:
  $$\left[\begin{array}{ccc} H & H & H \\ | & | & | \\ -C-C-C- \\ | & | & | \\ H & H & H \end{array} \right] \times$$

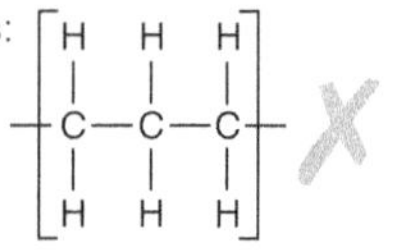

- **Addition polymers** are so useful because they are inert and **unreactive**. The main disadvantage of these plastic materials is they are almost completely **non-biodegradable**. Plastic refuse builds up in landfill sites; some plastics rot slowly to produce poisonous gases.

  One alternative is to **incinerate** waste plastic, using the heat to generate electricity. Chemists are working to limit the amount of **poisonous combustion products** being released, e.g. burning PVC evolves HCl fumes which add to acid rain.

  Waste plastics can be **recycled**, but they must be **sorted** because mixing plastics produces a soft, almost useless, product. Many waste plastics are cracked to make simple molecules which are useful feedstocks for the chemical industry.

  Biodegradable plastics are being produced that decompose naturally. Examples of biodegradable polymers are isoprene (2-methyl-1,3-butadiene), maize and starch. However maize and starch are food stuffs so making polymers out of these materials contributes to high world food prices.

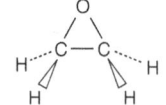

Fig. 8.3 Epoxyethane

- Occasionally you may meet **epoxyethane** which is made industrially by passing a mixture of air (or **oxygen**) and **ethene** over a **silver catalyst** (see Fig. 8.3). There is a danger of explosion because the ethene/oxygen mix is explosive. Epoxyethane is very reactive because the bond angle is 60°, which introduces bond strain caused by repulsion between displaced electrons. Epoxyethane is used to manufacture **epoxy resins**.

TESTS
RECALL TEST

1 What is required to convert methane into chloromethane?

   _____ (2)

2 How can the chloromethane be converted into dichloromethane?

   _____ (2)

3 What is a free radical?

   _____ (2)

4 Write equations for the free radical substitution of ethane by bromine. You should label the stages initiation, propagation, and termination.

   (6)

5 How are the components of crude oil separated?

   _____ (2)

6 Give two uses for the cracking of alkanes.

   _____ (2)

7 What mechanism is associated with the reactions of alkenes?

   _____ (1)

8 Generally what conditions are required when alkenes react?

   _____ (1)

9 Finish these equations:

   a $CH_2CH_2 + HBr \rightarrow$ _____

   b $CH_2CH_2 + Br_2(CCl_4) \rightarrow$ _____

   c $CH_2CH_2 + Br_2(aq) \rightarrow$ _____

   d $CH_2CH_2 + H_2O \rightarrow$ _____ (4)

Unit 8 — TESTS

10 How are alkenes converted into alcohols?

_____

_____ (1)

11 How could hexene be converted into hexane?

_____

_____

_____

_____ (3)

12 What is the test for alkenes?

_____

_____ (1)

13 Give the repeating unit for the polymer made when CH_3CHCH_2 polymerises.

_____

_____ (1)

14 What is the major disadvantage of polyalkanes?

_____

_____ (2)

(Total 30 marks)

CONCEPT TEST

1 Alkanes are a rich source of useful chemicals.

   a Give the reagents and conditions necessary to make methane into tetrachloromethane.

_____

_____ (2)

   b In the laboratory, how could ethene be converted into ethanol?

_____

_____ (2)

   c Industrially, tonnes of ethene are made into ethane-1,2-diol. How could this reaction be carried out on a small scale?

_____

_____

_____ (2)

   d How may octane be made from octene?

_____ (2)

2 a Some vegetable oils contain long-chain unsaturated molecules. Which chemical reaction would show that palm oil is saturated like animal fat, while sunflower oil is unsaturated? Give the reagent, conditions, and observations.

Reagent _____

Conditions _____

Observation with palm oil _____

Observation with sunflower oil _____ (4)

b Biodegradable polymers could be made from vegetable oil. Give one advantage and one disadvantage of biodegradable polymers.

_____

_____

_____ (2)

3 Free radicals are damaging to humans. The main sources are sunlight, smoke, and certain reactive chemicals.

a What is a free radical?

_____ (2)

b Explain how sunlight produces free radicals in your skin.

_____ (1)

c For a -CH_2- group on a molecule, show how it can be converted into -CHCl- by chlorine and sunlight.

_____

_____

_____

_____ (3)

(Total 20 marks)

Unit 9 HALOALKANES

> You will learn faster if you draw the mechanisms of these reactions.

- **Haloalkanes** (halogenoalkanes) consist of alkane molecules that have one or more hydrogen atoms replaced by **halogens** (F, Cl, Br, I). An example is bromoethane CH_3CH_2Br (see Fig. 9.1).

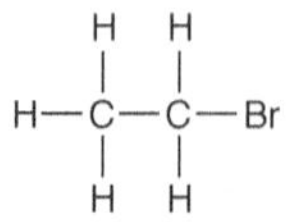

Fig. 9.1 Bromoethane

- Halogen atoms have greater numbers of electrons than hydrogen atoms, increasing the **induced Van der Waals forces**. The covalent C-halogen bond is polar (with the exception of C-I), producing **dipole–dipole interactions** between molecules. These intermolecular forces cause halogenoalkanes to have higher boiling points than the corresponding alkanes.

- Haloalkanes usually react by **nucleophilic substitution** by species such as OH^-, NH_3, and CN^- (see Fig. 9.2 and unit 8).

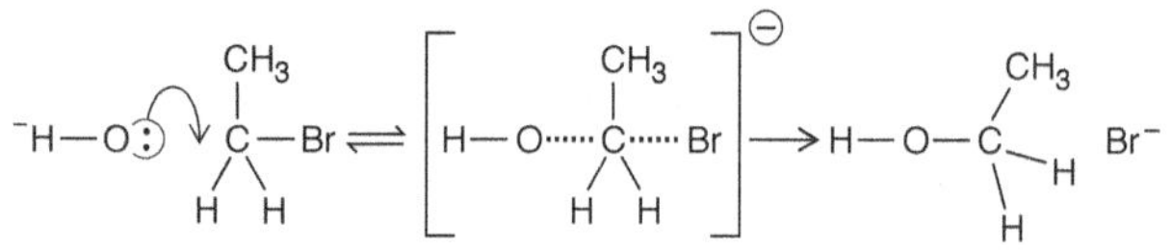

Fig. 9.2 Nucleophilic substitution

- Note that the **conditions** for these reactions include **heating under reflux** (see Fig. 9.3).

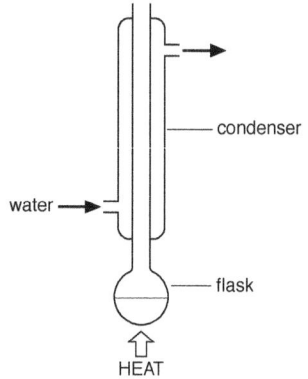

Fig. 9.3 Heat under reflux

Bond	Enthalpy (kJ mol^{-1})
F - C	485
Cl - C	340
Br - C	275
I - C	240

- Haloalkanes are reactive due to their weak bond enthalpies and because their bonds are easily polarised. The trend in reactivity of the haloalkanes is best explained using the bond enthalpies (see left). F-C is the strongest bond so the least reactive. I-C is the weakest bond so the most reactive.

  As the bonds become progressively weaker from F-C, to C-Cl, to C-Br, to C-I, the molecules become more reactive, so they react faster.

- **Aqueous hydroxide** ions from NaOH act as a nucleophile and substitute the Br atom to form an **alcohol**. This reaction is often referred to as **alkaline hydrolysis**.

  $CH_3CH_2Br + KOH(aq) \rightarrow CH_3CH_2OH + KBr$

- **Alcoholic hydroxide** ions from KOH dissolved in dry ethanol convert a haloalkane into an **alkene** by **elimination**. (See Fig. 9.4.)

  $CH_3CH_2Br + KOH(alcohol) \rightarrow CH_2CH_2 + KBr + H_2O$

  The OH⁻ ions are unable to act as nucleophiles because they are attached by hydrogen bonding to the alcohol molecules. They can still act as a base, accepting protons H⁺. Removal of H⁺ from bromoethane causes a Br⁻ ion to also leave; the electron pair from the broken C-H bond then forms an alkene pi bond.

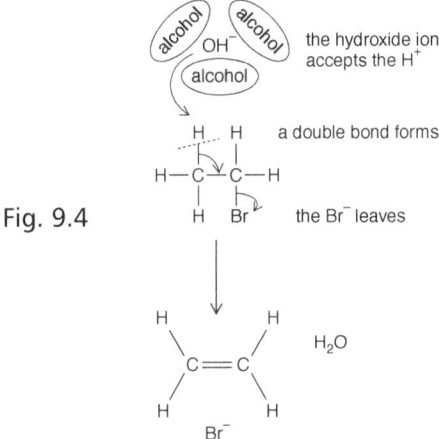

Fig. 9.4

- **Alcoholic ammonia** forms **amines** when heated with halogenoalkanes.

  $CH_3CH_2Br + 2NH_3(alcohol) \rightarrow CH_3CH_2NH_2 + NH_4Br$

  The reaction produces a low yield because the (**primary**) amine is also a nucleophile and will attack bromoethane to produce **secondary**, **tertiary**, and **quaternary** amines.

  $CH_3CH_2NH_2 \rightarrow (CH_3CH_2)_2NH \rightarrow (CH_3CH_2)_3N \rightarrow (CH_3CH_2)_4N^+$

- **Alcoholic cyanide** ions form **nitriles**.

  $CH_3CH_2Br + KCN(alcohol) \rightarrow CH_3CH_2CN + KBr$

  This reaction **increases** the carbon chain by **one C atom**.

- The **test** for a haloalkane is to add aqueous acidified **silver nitrate**. The water acts as a nucleophile and slowly substitutes OH for the halogen.

  R-Hal + H_2O → R-OH + HHal

  The halide ion released then combines with the silver ions to form a **precipitate**.

  $Ag^+ + Hal^- \rightarrow AgHal$

  The precipitate colours are:

  iodoalkane → **pale yellow** (AgI);

  bromoalkane → **off white** (AgBr);

  chloroalkane → **white** (AgCl).

- The reactivity of halogenoalkanes make them useful **intermediates in industry**. The halogen atoms are generally easily added to molecules, easily substituted, and haloalkanes are easily converted into alkenes.

Unit 9 — HALOALKANES

- The relatively strong C-Cl bond in **organo-chlorine** compounds such as insecticides makes them stable and **persistent**.

- **Chlorofluorocarbons** (CFCs) are compounds in which some or all of the hydrogen atoms are replaced by chlorine and fluorine atoms.

> A **volatile** liquid is one that evaporates easily. It does not mean that it is unreactive.
>
> A chemical that is unreactive is said to be **inert**.

CFCs are useful because they are **inert, volatile, and non-toxic**. They are used as refrigerants (in refidgerators and in air conditioning units), aerosol propellants, plastics foaming agents, in dry cleaning, and for degreasing metal.

- The **stability** of CFCs allows them to survive into the **ozone layer**, where UV radiation causes the C-Cl bonds to break.

Ozone (O_3) is continually being formed high in the atmosphere by this reversible reaction:

$$O_2 + O \rightleftharpoons O_3$$

The ozone layer is useful because it blocks harmful ultra-violet light from reaching the Earth's surface where it would otherwise cause increased skin cancers and lowered crop yields.

The CFC molecules, being inert and volatile, diffuse to the ozone layer where the ultra-violet light decomposes them to make chloro free radicals (which are chlorine atoms, Cl•).

The chloro free radicals break down the ozone molecules in two steps:

$$O_3 + Cl^\bullet \rightarrow O_2 + OCl^\bullet$$

$$OCl^\bullet + O_3 \rightleftharpoons 2O_2 + Cl^\bullet$$

Note that the chloro free radical is regenerated, so it is a catalyst. Each chloro free radical is able to decompose thousands of ozone molecules.

Chemists supported the ban on the use of CFCs and researched alternatives. Modern non-toxic and non-flammable **alternatives** are the more expensive **hydrofluorocarbons** (HFCs) such as CH_2FCF_3, which produce almost zero ozone depletion. Another alternative is carbon dioxide which is used as a blowing agent for expanded polymers.

This is a good example of **how science works**. Scientists detected that the ozone layer was being depleted, provided important evidence that enabled international action to be taken to reduce and phase out the use of CFCs, and researched safer ozone-friendly alternatives. Some experts state that it may take many decades before the ozone layer is repaired.

TESTS
RECALL TEST

1 Name the mechanism which predominates in the haloalkane reactions.

_____ (2)

2 Finish these equations.

   a $CH_3CH_2Cl(l) + NaOH(aq) \rightarrow$ _____

   b $CH_3CH_2I(l) + NaOH(ethanol) \rightarrow$ _____

   c $CH_3CH_2Br(l) + KCN(ethanol) \rightarrow$ _____

   d $CH_3CH_2Br(l) + KOH(ethanol) \rightarrow$ _____ (8)

3 Identify the product formed when CH_3CH_2Br is heated under reflux with aqueous potassium hydroxide.

_____

_____ (2)

4 State the reagents and conditions required to convert CH_3Cl into CH_3CN.

_____

_____ (2)

5 Hydroxide ions react with bromoethane in two different ways, depending on the conditions. State the two conditions and the two organic products.

_____

_____

_____

_____ (4)

6 Aqueous silver nitrate is mixed with an organic compound to produce a cream precipitate. Identify the cream precipitate and the functional group in the original organic compound.

_____

_____ (2)

7 Why are halogenoalkanes immiscible (insoluble) in water soluble in water?

_____

_____ (2)

8 Why does chloroethane have a much higher boiling point than propane?

_____

_____ (2)

Unit 9 — TESTS

9 Name these haloalkanes:

a CH₃F

b CH₂CH₂Br

c CH₂BrCH₂Br

d CH₃CHClCHClCH₂Br   (8)

10 Explain how CFCs are destroying the ozone layer by aswering these questions.

a What is the formula of ozone?

b What are CFCs?

c Write an equation to show how ozone forms naturally.

d What condition is require to change CFC molecules into free radicals?

e Write two equations to show how ozone is destroyed by chloro free radicals.

f What are the alternatives to CFCs?

(8)

(Total 40 marks)

CONCEPT TEST

1 Give examples of bromo compounds with a molecular formula C_4H_9Br:

a a primary haloalkane with a branching side chain,

(1)

b a secondary haloalkane.

(1)

2 a 2-bromopropane will react in two ways with KOH depending on the conditions. State the two conditions and name the organic products.

Condition 1 _____ Product 1 _____ (2)

Condition 2 _____ Product 2 _____ (2)

b Both **i** 2-aminopropane and **ii** 2-methylpropanenitrile may be formed from 2-bromopropane. Give the reagents and conditions.

  i Formation of 2-aminopropane:

  Reagents _____ Conditions _____ (2)

  ii Formation of 2-methylpropanenitrile:

  Reagents _____ Conditions _____ (2)

c How could you show in the laboratory that a compound contained a C-Br group?

_____ (2)

3 Give the structural formula of an isomer of $C_4H_{10}Br$ which is:

a a primary haloalkane _____

b a secondary haloalkane _____

c a tertiary haloalkane _____ (3)

4 Here are three reactions of 1,2-dibromopropane:

$$CH_3CHBrCH_2Br$$

↓ A   ↓ B   ↓ C

$CH_2CH(OH)CH_2OH$ $CH_3CH=CH_2$ $CH_3CH(CN)CH_2CN$

a Give the reagents and conditions to convert propan-2-ol for the reactions A to C:

Reaction A: Reagents _____ Conditions _____

Reaction B: Reagents _____ Conditions _____

Reaction C: Reagents _____ Conditions _____ (6)

b 2-bromopropane will also form an amine with ethanolic ammonia. Give the structural formula of this product.

_____
_____
_____ (2)

c 2,2-dimethyl 1-chloropropane reacts differently to 2-chloropropane. Identify the products when 2,2-dimethyl 1-chloropropane reacts with the following. If the reagent does not react, then state that it does not react.

  i KOH(ethanol) _____
  _____
  _____

  ii KOH(aq) _____ (2)

(Total 25 marks)

Unit 10: ALCOHOLS

> You need to recall the **four reaction types** of alcohols –
>
> **halogenation**, **oxidation**, **dehydration**, and **esterification**.

- **Alcohols** consist of a **hydroxyl group** -OH covalently bonded to a hydrocarbon. Ethanol CH_3CH_2OH (see Fig. 10.1) is an example of an aliphatic alcohol, in which -OH groups are bonded to straight or branched chain hydrocarbons.

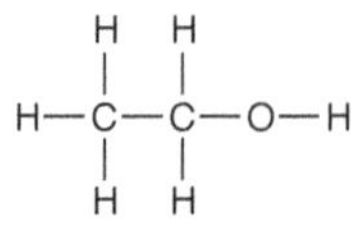

Fig. 10.1 Ethanol

The O-H bond is **polar** $O^{\delta-}$—$H^{\delta+}$, which causes **hydrogen bonding**. Alcohols have much higher boiling points than the corresponding alkanes e.g. ethanol (M_r 46) 76-°C; propane (M_r 44) b.p. –42-°C. The H-bonds also enable alcohols to **dissolve** in water (solubility decreases as the non-polar hydrocarbon chain increases in size).

- In industry phosphoric acid and steam, using high temperature and pressure, are used to produce **ethanol**:

  $CH_2CH_2 + H_2O \rightarrow CH_3CH_2OH$. (using H_3PO_4, 300 °C and 70 atm)

  Ethanol is also produced by **fermentation**. A source of sugar (e.g glucose) is mixed with water, yeast is added, the mixture kept warm, and air is excluded. This is how alcoholic drinks are made. The mixture may be distilled to produce pure ethanol which may be used as a fuel.

  The equation for fermentation of glucose is:

  $C_6H_{12}O_6 \rightarrow 2CH_3CH_2OH + 2CO_2$

- Alcohols are used as solvents, for example, ethanol is used to dissolve perfumes. A useful solvent is methylated spirit which is a mixture of ethanol and poisonous methanol.

  Ethanol, made from fermentation and purified by distillation, is used as a fuel. The **biofuel** ethanol, when made from sugar cane, is said to be **carbon neutral** because there is no net addition to atmospheric carbon dioxide when the fuel is burnt.

  Methanol is added to petrol to improve combustion.

> A **biofuel** is a fuel that is made from recent biological sources, such as plants, algae or animal sources. Ethanol is a biofuel is made from sugar cane, but is not a biofuel if made from crude oil.
>
> An increased use of biofuels have helped to increase world food prices. **Example :** Grain used to make ethanol, and palm oil used to make bio-diesel are a particualr problem.
>
> Research is being focussed on biofuels that do not use agricultural land. **Example:** algae grown on brackish (salty) water, and special grasses grown on infertile land.

- You should know the difference between **primary**, **secondary**, and **tertiary** alcohols (see Fig. 10.2).

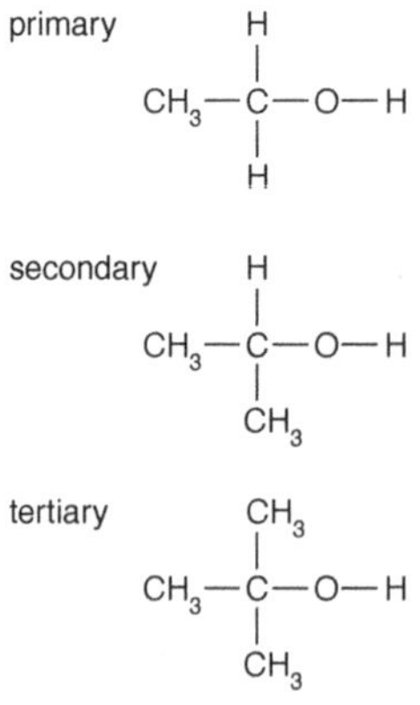

Fig. 10.2

- **Halogenation** involves nucleophilic substitution to convert alcohols into the corresponding halogeno compound, e.g. ethanol to the **haloalkanes** chloroethane, bromoethane, and iodoethane:

    $CH_3CH_2OH + HCl \rightarrow CH_3CH_2Cl + H_2O$ (heat under reflux)

    $CH_3CH_2OH + PCl_5 \rightarrow CH_3CH_2Cl + POCl_3 + HCl$ (room temperature)

    $CH_3CH_2OH + HBr \rightarrow CH_3CH_2Br + H_2O$
    (heat under reflux, with HBr made in situ using NaBr and conc. H_2SO_4)

    $CH_3CH_2OH + PI_3 \rightarrow CH_3CH_2I + PIO + HI$
    (room temperature with PI_3 made in situ by mixing iodine with red phosphorus)

> For most reactions you write 'heat under reflux' for the conditions (see Fig 8.3), but some reactions do occur at room temperature.

> Converting alcohols into halogenoalkanes is useful in organic synthesis because you can convert them into many other types of compound.

Converting alcohols into halogenoalkanes is useful in organic synthesis because you can convert them into many other types of compound.

> To **test** for an alcohol, suggest adding **PCl_5**. Dense white fumes (of HCl) show an OH group is present. NB also gives white fumes with water and carboxylic acids.

- Alcohols may be **oxidised** by combustion. Ethanol burns with a clean blue flame. $CH_3CH_2OH + 3O_2 \rightarrow 2CO_2 + 3H_2O$

    When exposed to the air, **primary alcohols** RCH_2OH will **oxidise** very slowly to **aldehydes** RCHO and then to carboxylic acids RCOOH (e.g. beer and wine slowly change to vinegar). The strong oxidising agents acidified potassium dichromate(VI) $K_2Cr_2O_7$ and acidified potassium permanganate manganate(VII) $KMnO_4$ are used in the lab.

- To stop the oxidation at the **aldehyde**, the oxidising agent $K_2Cr_2O_7$ with H_2SO_4 is dripped into hot **primary alcohol**, and the aldehyde is **distilled** off as it forms.

    $CH_3CH_2OH + [O] \rightarrow CH_3CHO + H_2O$

    The aldehyde has a polar C=O so has a lower b.p. (49-°C) than the hydrogen-bonded alcohol (76-°C). Use $NaBH_4(aq)$ to reduce back to the alcohol.

    $CH_3CHO + 2[H] \rightarrow CH_3CH_2OH$

    To oxidise a primary alcohol to the carboxylic acid, heat under reflux with the oxidising agent and then distil off the product.

    $CH_3CH_2OH + 2[O] \rightarrow CH_3COOH + H_2O$
    (Use $LiAlH_4$ to reduce back to the alcohol.)

> Notice you just write [O] to show there is an oxidising agent.

- **Secondary alcohols** oxidise to form a **ketone** e.g. propan-2-ol plus oxidising agent form propanone and water.

    $CH_3CH(OH)CH_3 + 2[O] \rightarrow CH_3COCH_3 + H_2O$

- **Tertiary alcohols** cannot easily be oxidised except by combustion.

Unit 10 — ALCOHOLS

- All alcohols can be **dehydrated** to make **alkenes** (see Fig. 10.3). The reagents and conditions are to either heat under reflux with conc. H_2SO_4 or H_3PO_4 or to pass the alcohol vapour over hot pumice or Al_2O_3.

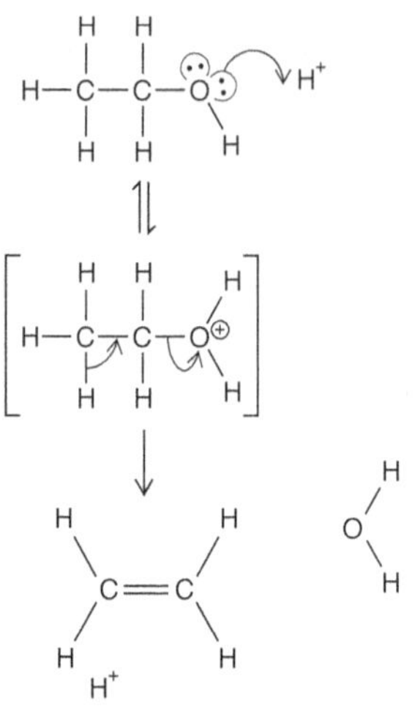

Fig. 10.3 Dehydration of alcohol

- All alcohols will join with carboxylic acids in condensation reactions to form **esters**. $CH_3CH_2OH + CH_3COOH \rightarrow CH_3COOCH_2CH_3 + H_2O$

  The reaction happens when the two compounds are mixed and warmed. Adding concentrated H_2SO_4 catalyses the reaction (increases the rate) and removes the water (increases the yield).

- **Infrared spectroscopy** is used to identify alcohols and to distinguish them from aldehydes, ketones, and carboxylic acids. It depends on bonds absorbing infrared radiation to increase their vibrational energy. Different bond types absorb at different frequencies. The **spectrum** shows percentage transmittance of energy (on the y-axis) against the frequency of the radiation (on the x-axis). The unit of frequency is the **wavenumber** cm^{-1} (waves per centimetre), chosen to give a convenient scale.

  Either you may be required to state the wave number that you would expect for molecule, or the examiner may give you the wave numbers either in a graph or stated within a question, and you must identify bond.

  Below are the characteristic wavenumbers for infra-red absorbtion:

Bond	Functional Group	Wavenumber/cm^{-1}	Intensity
C=O	aldehyde ketone carboxylic acid	1700-1740	S
C-O	alcohols	1150-1310	S
O-H	hydrogen bonded alcohol	3200-3550	S (broad)
O-H	not hydrogen bonded alcohol	3590-3650	S
O-H	carboxylic acid	2500-3300	M (broad)

Note: most molecules produce a broad peak around 3000 cm^{-1} due to C-H bonds.

S = sharp, M = medium, broad = the peak is broad.

RECALL TEST

1 Name the reaction types which are common in in the alcohol reactions.
_____
_____
_____ (4)

2 Why are short-chain alcohols soluble in water, while long-chain alcohols are insoluble?
_____
_____
_____ (3)

3 Why does ethanol have a much higher boiling point than ethanal?
_____
_____ (2)

4 Name these alcohols:

   a CH_3CH_2OH _____

   b $CH_3CHOHCH_3$ _____

   c $(CH_3)_3COH$ _____

   d $(CH_3)_3CCH_2OH$ _____ (8)

5 Fill in the rest of these equations.

   a $CH_3CH_2OH + Na \rightarrow$ _____

   b $CH_3CH_2OH + [O] \rightarrow$ _____ using $KMnO_4(aq)$ with $H_2SO_4(aq)$

   c $CH_3CH_2OH \rightarrow$ _____ using concentrated $H_2SO_4(l)$

   d $CH_3COOH + CH_3CH_2OH \rightarrow$ _____ with concentrated $H_2SO_4(l)$

   e $CH_3CH_2OH + PCl_5 \rightarrow$ _____

   f $CH_3CH_2OH + HBr \rightarrow$ _____

   g $CH_3CHOHCH_3 + [O] \rightarrow$ _____ (8)

(Total 25 marks)

Unit 10

TESTS

CONCEPT TEST

1 Give examples of alcohols with a molecular formula C_4H_9OH:

   a a primary alcohol with a branching side chain,

   _____

   _____ (1)

   b a secondary alcohol.

   _____

   _____ (1)

2 a Propan-1-ol will react in two ways with acidified potassium dichromate solution depending on the conditions. State the two conditions and name the organic products.

   Condition 1 _____ Product 1 _____ (2)

   Condition 2 _____ Product 2 _____ (2)

   b Both i sodium propan-2-oxide and ii propene may be formed from propan-2-ol. Give the reagents and conditions.

   i Formation of sodium propan-2-oxide:

   Reagents _____ Conditions _____ (2)

   ii Formation of propene:

   Reagents _____ Conditions _____ (2)

   c How could you show in the laboratory that a compound contained a C-O-H group?

   _____

   _____

   _____ (2)

3 Give the structural formula of an isomer of $C_4H_{10}O$ which is:

   a a primary alcohol

   b a secondary alcohol

   c a tertiary alcohol (3)

4 Here are three reactions of propan-2-ol:

$$\text{CH}_3\text{CHOHCH}_3$$

```
         A              B              C
         ↓              ↓              ↓
   CHCOOCH(CH₃)₂   CH₃CHBrCH₃    CH₃CH=CH₂
```

a Give the reagents and conditions to convert propan-2-ol for the reactions A to C:

Reaction A: Reagents _____ Conditions _____

Reaction B: Reagents _____ Conditions _____

Reaction C: Reagents _____ Conditions _____ (6)

b Propan-2-ol will also form an ester with ethanoic acid. Give the structural formula of this ester.

_____
_____
_____ (2)

c 2-methylpropan-2-ol reacts differently to propan-2-ol. Identify the products when 2-methylpropan-2-ol reacts with the following. If the reagent does not react with 2-methylpropan-2-ol, then state that it does not react.

  i heated aluminium oxide _____

  ii potassium dichromate _____ (2)

(Total 25 marks)

Unit 11 — ENERGETICS: ENTHALPY CHANGE

> You must remember that an enthalpy change is a **heat change** and **not** an energy change (which can involve doing work).

- The **heat** evolved from or absorbed by a reaction at constant pressure is called the **enthalpy change**.

  When a reaction gives out heat, we say that the reaction is **exothermic**. The heat change (units $J\,mol^{-1}$) is given a **negative sign** because the reacting chemicals have **lost heat** to their surroundings. Combustion of a piece of paper and respiration are examples of exothermic reactions.

  When a reaction takes in heat, we say that the reaction is **endothermic**. The heat change is given a **positive** sign because the reacting chemicals have **gained heat** from their surroundings. An example is photosynthesis.

> Enthalpy changes also occur during **physical changes** such as boiling and freezing.

- Enthalpy changes are measured for **1 mole** of substance under standard conditions (**298 K**, 1 atmosphere pressure, **101 325 Pa**). When you define a standard enthalpy change remember to state the standard conditions. **Example:** The standard enthalpy of vaporisation of water:

  $H_2O(l) \rightarrow H_2O(g) \quad \Delta H^\ominus_v = +41.1\ kJ\,mol^{-1}$

> Remember that you must know the definitions given in this spread: you cannot start to attempt calculations without them.

- You must recall the definitions (but not numerical values) of certain standard enthalpy changes.

  Enthalpy of formation: The enthalpy change when one mole of a substance is formed from its constituent elements in their normal states (under standard conditions). **Example:**

  $Na(s) + \tfrac{1}{2}Cl_2(g) \rightarrow NaCl(s) \quad \Delta H^\ominus_f (NaCl) = -411\ kJ\,mol^{-1}$

  Enthalpy of combustion: The enthalpy change when one mole of a substance is completely combusted in oxygen (under standard conditions). **Example:**

  $CH_4(g) + 2O_2(g) \rightarrow CO_2(g) + 2H_2O(l) \quad \Delta H^\ominus_c (methane) = -890.4\ kJ\,mol^{-1}$

  Enthalpy of neutralisation: The enthalpy change when one mole of water is formed from the reaction of an acid and a base (under standard conditions). **Example:**

  $NaOH(aq) + HCl(aq) \rightarrow NaCl(aq) + H_2O(l) \quad \Delta H^\ominus_n = -57.1\ kJ\,mol^{-1}$

  Enthalpy of reaction: The enthalpy change that accompanies a reaction between the amounts of substances (under standard conditions) shown in the balanced chemical equation. **Example:** The reaction between ammonia and fluorine.

  $NH_3(g) + 3F_2(g) \rightarrow 3HF(g) + NF_3(g) \quad \Delta H^\ominus_r = -875\ kJ\,mol^{-1}$

> The enthalpy change for **making** a bond has the same value but opposite sign.

  Mean bond enthalpy: The **average** energy required to **break** one mole of a particular kind of bond derived from a wide range of molecules that contain the bond. The environment of a given bond type may be different in different molecules. As a result, mean bond enthalpy values will not exactly agree with bond enthalpy values derived from one particular molecule. **Example:**

  Mean bond enthalpy$_{(C-H)} = +412\ kJ\,mol^{-1}$

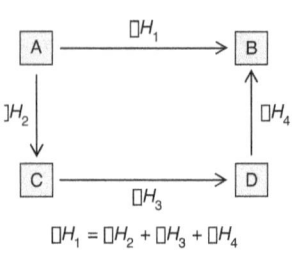

Fig. 11.1 Hess's law

- **Hess's law** states that the enthalpy change accompanying a reaction is **independent of the route** taken. Suppose A → B directly and also A → C → D → B indirectly. The enthalpy change accompanying A → B equals the sum of the enthalpy changes accompanying A → C, C → D, and D → B (see Fig. 11.1).

- To work out the heat change, use the relationship:

  quantity of heat = mass x specific heat capacity x temperature change

  q = mcΔT

  q = quantity of heat, m = mass, ΔT = temperature change

  c = specific heat capacity (The specific heat capacity of water is 4.18 $Jg^{-1}K^{-1}$)

  > Note the density of water = $1\,g\,cm^{-3}$.

  Example: An excess of NaOH solid, 5.00 grams, is added to 200.0 cm^3 of 0.500 mol dm^{-3} hydrochloric acid. This is how to calculate the enthalpy change for the reaction.

  The temperature rises from 20.0°C to 26.8°C.

  Thus the temperature change = 26.8 - 20.0 = 6.8°C = 6.8K.

  The heat change (q) = 200 x 4.18 x 6.8 = 5685 J.

  The number of moles of HCl   = concentration x volume (in dm^3)

      = 0.5 x 0.200 = 0.1 moles HCl.

  Enthalpy of reaction   = heat change (q) / moles

      = 5685 / 0.1 = 56850 $Jmol^{-1}$

  Note that the temperature rose so the reaction was exothermic, so the enthalpy of the reaction must have a negative value:

  Enthalpy of reaction = 56.9 kJ mol^{-1}

  Note that:

  (i) the mass of the NaOH was not included because it was the water that was heated up;

  (ii) heat was given off so the enthalpy ust have a negative sign,

  (iii) the moles were based on the HCl becasue the NaOH was in excess.

  (iv) The answer was given to 3 significant figures as the data given was to three figures.

- There are many ways of organising the Hess's law **calculations** you will meet at AS level. Here is a good way to do calculations that will produce fewer mistakes (see Figs 11.2, 11.3, and 11.4 on the next page).

  One: Translate all the Δ*H* terms given in the question into chemical equations.

  Two: Across the full width of the page, write out the chemical equation that corresponds to the enthalpy change you seek. This equation represents the **direct route**.

  Three: There will be an **indirect route** between the reactants and the products written in step **two**. Inspect the data you are given to see if there are obvious intermediate substances. If you are given ΔH_f, then **intermediates** might be elements; if ΔH_c (for hydrocarbons), then intermediates could be CO_2 and H_2O.

  Four: Set out the Hess's law **enthalpy** cycle. Make sure the **arrows** point in the **correct directions**.

  Example: Direct route: elements → combustion products; indirect route: elements → compound → combustion products.

  Remember to

  (i) write the **values** for enthalpy changes over the arrows;

  (ii) **reverse the signs** of the enthalpy changes if the direction of the indirect enthalpy change is in the opposite direction from the direct change. This is shown in the examples;

  (iii) multiply molar enthalpy changes according to the number of moles of substance in the equations, other word **balance** the equations and values.

  Example: If two H_2O molecules are made from the combustion of hydrogen, then to **balance** the values you must write the total enthalpy change as 2 × $\Delta H^{\ominus}{}_{c(H_2)}$.

Unit 11 — ENERGETICS: ENTHALPY CHANGE

> If you have found a way that works for you AND you get correct answers most of the time, then stick with it.

Five: Draw an arrow from the starting substances to the end products via the intermediate reaction(s). **Sum** the enthalpy changes for the indirect route and **equate** these to the enthalpy change for the direct route.

Check that you have **balanced** the equations and used them correctly to multiply the values of the ΔH terms.

Check you have drawn the arrows in the **correct direction.**

Check that the signs of the enthalpy change terms are correct for the direction of the change concerned.

Fig. 11.2

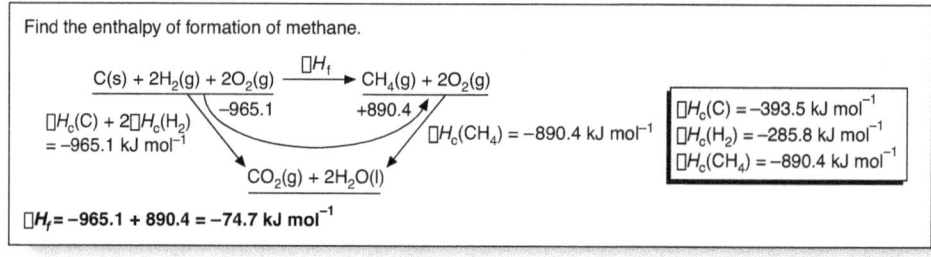

Find the enthalpy of formation of methane.

$C(s) + 2H_2(g) + 2O_2(g) \xrightarrow{\Delta H_f} CH_4(g) + 2O_2(g)$

$\Delta H_c(C) + 2\Delta H_c(H_2) = -965.1$ kJ mol^{-1}; $+890.4$

$CO_2(g) + 2H_2O(l)$

$\Delta H_c(C) = -393.5$ kJ mol^{-1}
$\Delta H_c(H_2) = -285.8$ kJ mol^{-1}
$\Delta H_c(CH_4) = -890.4$ kJ mol^{-1}

$\Delta H_f = -965.1 + 890.4 = -74.7$ kJ mol^{-1}

Fig. 11.3

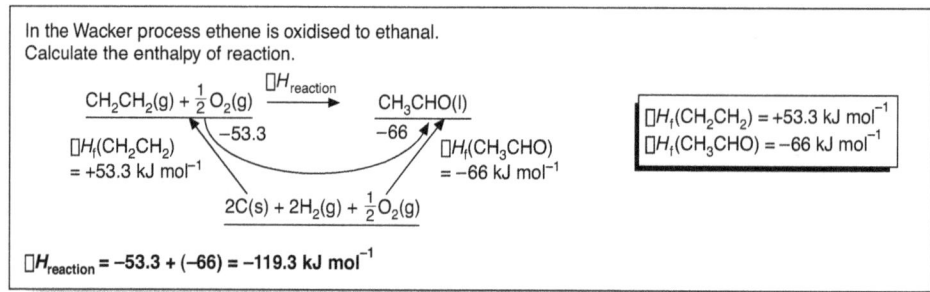

In the Wacker process ethene is oxidised to ethanal. Calculate the enthalpy of reaction.

$CH_2CH_2(g) + \tfrac{1}{2}O_2(g) \xrightarrow{\Delta H_{reaction}} CH_3CHO(l)$

$\Delta H_f(CH_2CH_2) = +53.3$ kJ mol^{-1}; -53.3; -66; $\Delta H_f(CH_3CHO) = -66$ kJ mol^{-1}

$2C(s) + 2H_2(g) + \tfrac{1}{2}O_2(g)$

$\Delta H_{reaction} = -53.3 + (-66) = -119.3$ kJ mol^{-1}

Fig. 11.4

Calculate the enthalpy of oxidation of ethanol to ethanoic acid.

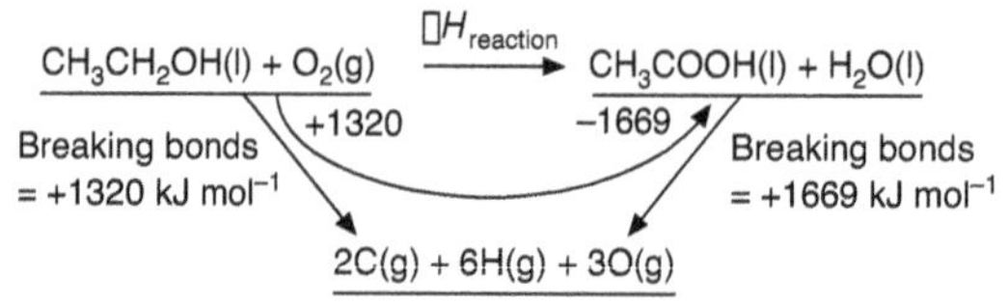

$CH_3CH_2OH(l) + O_2(g) \xrightarrow{\Delta H_{reaction}} CH_3COOH(l) + H_2O(l)$

Breaking bonds $= +1320$ kJ mol^{-1} ; $+1320$; -1669 ; Breaking bonds $= +1669$ kJ mol^{-1}

$2C(g) + 6H(g) + 3O(g)$

Effectively breaking bonds
$2 \times (C-H) = 2 \times (+412) = +824$
$1 \times (O=O) = +496$
Total $= +1320$
(Breaking bonds is endothermic)

Effectively making bonds
$1 \times (C=O) = -743$
$2 \times (O-H) = -2 \times (+463) = -926$
Total $= -1669$
(Making bonds is exothermic)

Bond enthalpies (kJ mol^{-1})
(C—C) +348
(C—H) +412
(C—O) +360
(C=O) +743
(O—H) +463
(O=O) +496

Enthalpy of reaction $= +1320 - 1669 = -349$ kJ mol^{-1}

Fig. 11.5 The enthalpies in short

ΔH_f: elements → compound

ΔH_c: compound + O_2 → combustion products (often CO_2, H_2O)

$\Delta H_{neutralisation}$: acid + base → water (+ a salt)

ΔH_{bond}: gaseous molecule → gaseous atoms

$\Delta H_{solution}$: solid + (aq) → aqueous ions

1st ionisation energy: gaseous atom → gaseous (+) ion

1st electron affinity: gaseous atom → (−) gaseous ion

TESTS
RECALL TEST

1. What sign does an enthalpy have in an exothermic reaction?
   _____ (1)

2. Define the enthalpy of:

   a formation

   _____
   _____
   _____ (3)

   b combustion

   _____
   _____
   _____ (3)

   c neutralisation

   _____
   _____
   _____ (3)

3. Define 'bond enthalpy'.
   _____
   _____
   _____ (3)

4. Define 'mean bond enthalpy'.
   _____
   _____
   _____ (3)

5. State Hess's law.
   _____
   _____
   _____ (2)

6. Write the equation used to calculate the heat involved in changing the temperature of a mass of water.
   _____
   _____
   _____ (2)

(Total 20 marks)

Unit 11 — TESTS

Sometimes the **examiners** will give you data in a **previous** part of the question; sometimes they will give you **surplus** data.

CONCEPT TEST

1 As crude oil is going to run out, research has focussed on making organic compounds from coal. One idea is to react coal with water to make carbon monoxide which then reacts with hydrogen to make methanol.

Reaction A: $C(s) + H_2O(l) \rightarrow CO(g) + H_2(g)$

$\Delta H_f(CO_2(g)) = -394 \text{ kJ mol}^{-1}$

$\Delta H_f(CO) = -111 \text{ kJ mol}^{-1}$

$\Delta H_f(H_2O(l)) = -286 \text{ kJ mol}^{-1}$

$\Delta H_c(CO) = -283.0 \text{ kJ mol}^{-1}$

$\Delta H_c(CH_3OH) = -715.0 \text{ kJ mol}^{-1}$

a Use the data above to calculate the enthalpy of the reaction A.

(4)

b An alternative idea is to convert coal and water in the presence of hydrogen directly to methanol (but a catalyst has yet to be perfected).

Reaction B: $C(s) + H_2O(l) + H_2(g) \rightarrow CH_3OH(l)$

Calculate the enthalpy of reaction B.

(4)

c The methanol could then be converted to methanal, CH_2O.

Here are some bond enthalpies, in kJ mol^{-1}.

C-C	C-H	C-O	C=O	O-H	O=O
348	412	360	743	463	496

Calculate the enthalpy of reaction for the oxidation of methanol:

$CH_3OH(l) + \tfrac{1}{2}O_2(g) \rightarrow CH_2O(g) + H_2O(l)$

(3)

2 Given these enthalpies, calculate the enthalpy of formation of dinitrogen tetraoxide, N$_2$O$_4$:

$$2NO_2(g) \rightarrow N_2O_4(g) \quad \Delta H_{reaction} = -58.1 \text{ kJ mol}^{-1}$$

$$\tfrac{1}{2}N_2(g) + \tfrac{1}{2}O_2(g) \rightarrow NO(g) \quad \Delta H_f(NO) = +90.4 \text{ kJ mol}^{-1}$$

$$NO(g) + \tfrac{1}{2}O_2(g) \rightarrow NO_2(g) \quad \Delta H_{reaction} = -56.5 \text{ kJ mol}^{-1}$$

(2)

3 The oxidation of ammonia to make nitrogen monoxide is very important as the nitrogen monoxide may then be converted easily into nitric acid. The stoichiometric equation could be written as

$$4NH_3(g) + 5O_2(g) \rightarrow 4NO(g) + 6H_2O(l)$$

$$\Delta H_f(NH_3) = -46.2 \text{ kJ mol}^{-1}$$

$$\Delta H_f(NO) = +90.4 \text{ kJ mol}^{-1}$$

$$\Delta H_f(H_2O) = -286 \text{ kJ mol}^{-1}$$

a Use Hess's law to calculate the enthalpy of reaction for the oxidation of ammonia.

(3)

b Calculate the (Si-Cl) bond enthalpy in SiCl$_4$, given these values:

$\Delta H_f(SiCl_4(l)) = -640 \text{ kJ mol}^{-1}$

$\Delta H_a(Si) = +439 \text{ kJ mol}^{-1}$

(Cl-Cl) bond enthalpy = $+242 \text{ kJ mol}^{-1}$

(4)

(Total 20 marks)

Unit 12 — KINETICS: RATE

- **Rate** is the measure of how fast the **concentration** of a reactant or a product **changes** over time. The units of rate are usually **mol dm^{-3} s^{-1}**.

  The **factors** that **control** rate are temperature, concentration, pressure, catalyst, surface area, light.

- You must remember the **three requirements** for chemical reactions to successfully take place: reactants (atoms, molecules, or ions) must **collide** with each other with the **correct orientation** and with **sufficient energy**.

  Increasing the reactant concentration in solutions increases the reaction rate because there are more reactant particles in a given volume, so they **collide** and react **more frequently**.

  Increasing the pressure of reacting gases increases their concentrations. The reaction rate increases because there are more molecules in a given volume, so they **collide** and react **more frequently**.

  Increasing the surface area of a reactant (e.g. powdering a solid) increases the reaction rate as there is **greater contact** between the reactants, so more collisions take place per second.

- The **energy changes** that happen during a reaction may be expressed as a **reaction profile** (see Figs 12.1 and 12.2).

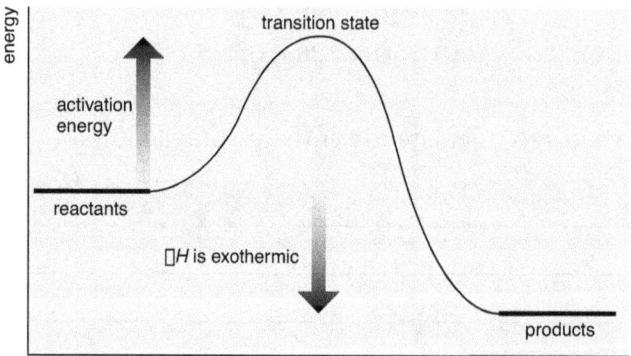

Fig. 12.1 Reaction profile for an exothermic reaction

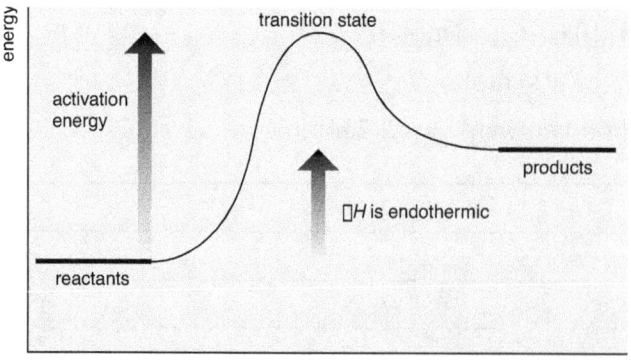

Fig. 12.2 Reaction profile for an endothermic reaction

- **Activation energy** is the minimum energy required in a collision for the particles to react. Units are J mol^{-1} or kJ mol^{-1} (just like enthalpy). It acts as an **energy barrier** that reactants must overcome if they are to react.

- When molecules react, they form **unstable groups** of atoms called **transition states**. A **trough** at the peak of a reaction profile indicates the existence of an unstable **intermediate compound** (see Fig. 12.3).

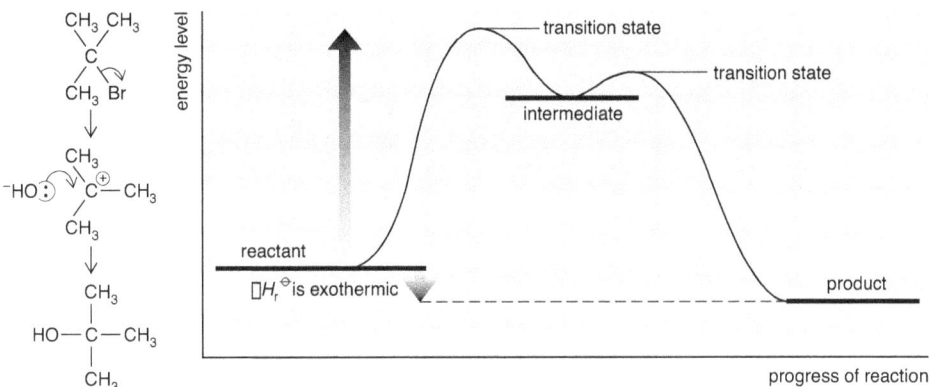

Fig. 12.3 An intermediate forms when a tertiary haloalkane undergoes nucleophilic substitution (S_N1).

In some reactions, the rate increases when reactants absorb energy from **light** so they can overcome the activation energy barrier.

- The **Maxwell–Boltzmann distribution** of molecular energies gives an instantaneous 'snapshot' (at a specified temperature) of the **proportion** of molecules in a sample that have a given energy (see Fig 12.4).

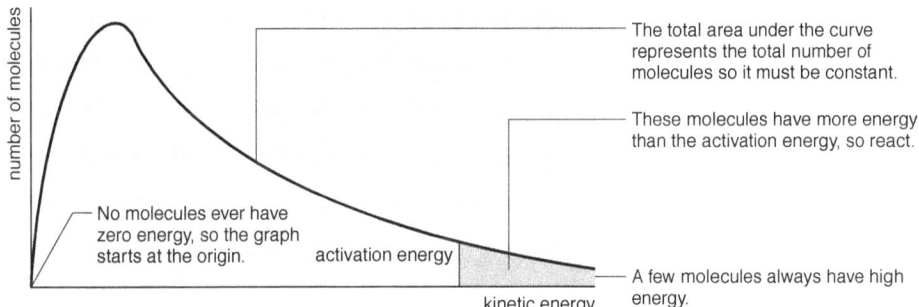

Fig. 12.4

Note that the molecules with the highest kinetic energy are on the right of the diagram. Look at the shaded region to the right of the vertical line representing the activation energy. You must understand that all the molecules in this area have sufficient kinetic energy to react on collision.

- **Increasing the temperature** increases the reaction rate because molecular **kinetic energy increases** as the temperature rises, which increases the number of molecules that collide with energies greater than the activation energy. Increasing the temperature also increases the **number** of collisions per second (see Fig. 12.5).

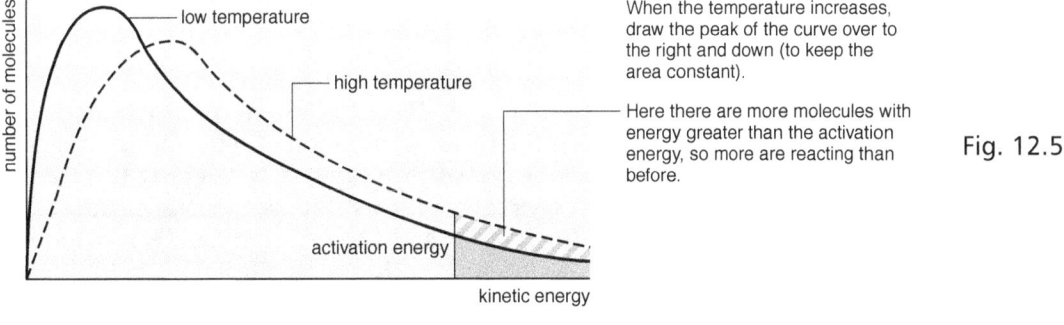

Fig. 12.5

Note that even a small increase in temperature can cause a large increase in rate.

Unit 12 — KINETICS: RATE

- **Catalysts** increase the reaction rate without themselves being used up. They function by creating **alternative** reaction routes (via intermediates) with **lower activation energies** (see Fig. 12.6).

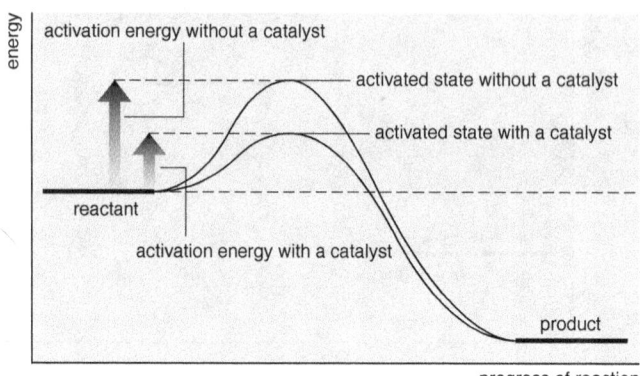

Fig. 12.6 Reaction profile of a reaction with and without a catalyst

As a result, there are more molecules present at a given temperature that have sufficient energy to react (see Fig. 12.7).

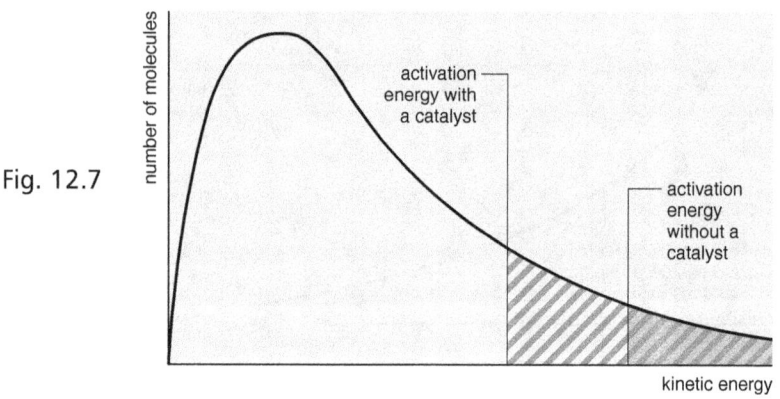

Fig. 12.7

- Catalytic **poisons** bind **irreversibly** (permanently) with catalysts and stop them working. **Example: Haem** is a porphyrin complex similar in structure to porphyrin synthetic dyes. This stores oxygen in red blood cells. It is poisoned by CO. (See Fig. 14.1)

- **Homogeneous catalysts** are in the **same** phase as the reactants.
  Example:-Fe^{2+}(aq) catalyses the redox reaction between peroxodisulfate and iodide ions:
  $$2I^-(aq) + S_2O_8^{2-}(aq) \rightarrow I_2(aq) + 2SO_4^{2-}(aq)$$

- **Heterogeneous catalysts** are in a **different** phase to the reactants.
  Example: Nickel catalysing the reaction between ethene and hydrogen to make ethane. The nickel acts as a surface catalyst, **adsorbing** the reactant gas molecules so they are **correctly aligned** for easy reaction, and then **desorbing** the product. Tungsten adsorbs too strongly and silver too weakly, so they are not suitable. Nickel and platinum are ideal.

TESTS
RECALL TEST

1 Define 'rate'. What are the units of rate?

   (2)

2 State six factors which influence rate.

   (6)

3 What three conditions are required for molecules to react?

   (3)

4 How does increasing concentration increase rate?

   (2)

5 How does increasing surface area increase rate?

   (2)

6 Draw an energy profile for an endothermic reaction with an intermediate. Label the intermediate, and any transition states.

   (2)

7 Define 'activation energy'.

   (3)

8 Draw a graph to show why increasing temperature increases rate.

   (3)

Unit 12 — TESTS

9 Explain why increasing temperature increases rate.

_____

_____

(2)

10 What is a catalyst?

_____

(1)

11 In general, how does a catalyst increase rate?

_____

_____

(2)

12 Give two specific ways in which it does this.

_____

_____

(2)

(Total 30 marks)

CONCEPT TEST

1 a Define 'activation energy'.

_____

_____

_____

(3)

b Draw and label the reaction profile for a reaction which is endothermic and forms an intermediate.

(3)

c Draw and label a graph to explain why rate increases when temperature increases.

(3)

d Explain in words why the rate increases when the temperature is increased.

_____

_____

_____

(3)

e What term is used to describe the time taken for the reactant concentration to change?
_____ (1)

2 a Draw a reaction profile to explain the effect of adding a catalyst.

(2)

b Explain in words why the rate increases when a catalyst is added.
_____
_____
_____
_____ (3)

c Solid iron is used to catalyst the reaction of nitrogen with hydrogen to produce ammonia. Is the catalyst heterogeneous or homogeneous? Explain your reasoning.
_____
_____
_____ (2)

(Total 20 marks)

Unit 13

EQUILIBRIUM

The reaction mixture is a **closed system**, which means that there must not be a loss or gain of reactants or products or energy.

- A **dynamic equilibrium** develops during all chemical reactions: **concentrations** of substances remain **constant** as reactants change into products (the **forward reaction**) and products revert to reactants (the **backward reaction** – or **reverse reaction**).

  At **equilibrium**, the **rates** of the forward and backward reactions are equal (see Fig. 13.1).

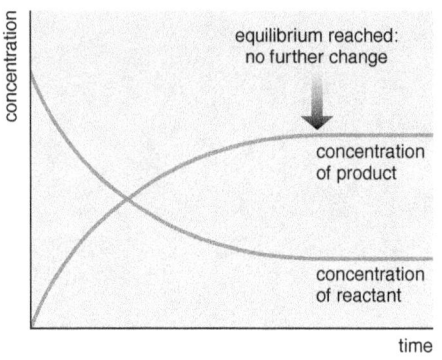 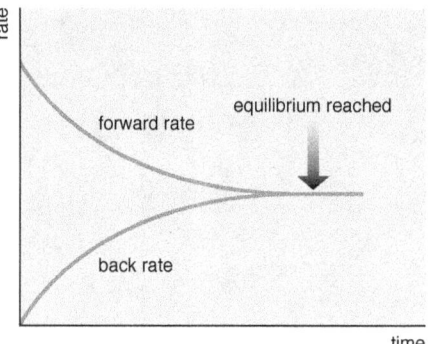

Fig. 13.1

The term **reversible reaction** is used to refer to equilibrium mixtures that contain significant amounts of both reactants and products.

Example: The production of ammonia NH_3 from hydrogen and nitrogen in the Haber process.

$N_2(g) + 3H_2(g) \rightleftharpoons 2NH_3(g)$

At equilibrium, N_2 and H_2 are reacting to make NH_3 at the same rate as NH_3 is decomposing to make N_2 and H_2.

- When **product** concentration at equilibrium is **large** compared with reactant concentration, then the equilibrium is said to **lie to the right** and the reaction goes **to completion**. When product concentration at equilibrium is very **small**, then the equilibrium is said to **lie to the left** and the reaction effectively does not take place.

- **Yield** is the percentage proportion of the product in the equilibrium mixture.

- When the **forward** reaction is **exothermic**, the **backward** reaction is **endothermic**. Example: The Haber process equilibrium consists of the forward reaction

  $N_2(g) + 3H_2(g) \rightarrow 2NH_3(g) \quad \Delta H^\ominus = -92 \text{ kJ mol}^{-1}$

  and the backward reaction

  $2NH_3(g) \rightarrow N_2(g) + 3H_2(g) \quad \Delta H^\ominus = +92 \text{ kJ mol}^{-1}$.

  The equilibrium reaction is written as

  $N_2(g) + 3H_2(g) \rightleftharpoons 2NH_3(g)$

  and is referred to as an **exothermic reversible reaction**.

- **Le Chatelier's principle** states that if the conditions of a system at equilibrium are changed then the equilibrium position will **shift** to resist the change. You may need to recall this definition, but it is more important to understand how to apply the idea.

If the **concentration** of a substance is **increased**, then the equilibrium will shift to **decrease** its concentration.

Example: If extra N_2 is added to an equilibrium mixture of H_2, N_2, and NH_3, then the equilibrium shifts to increase the concentration of NH_3 and decrease the concentration of N_2 (and H_2) (see Fig. 13.2).

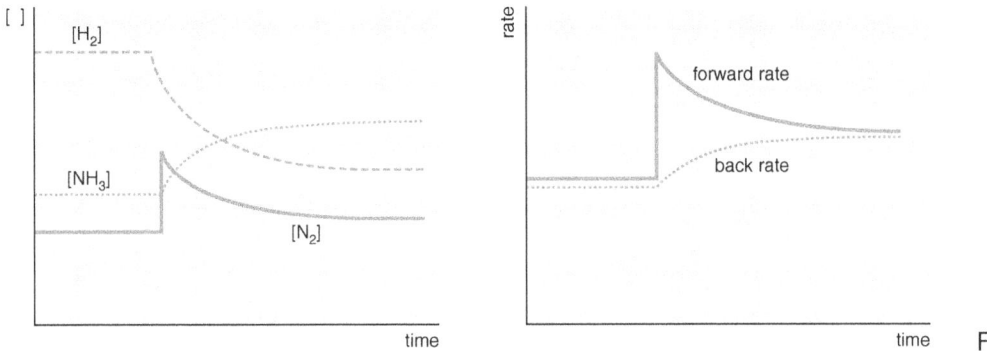

Fig. 13.2

If a reversible reaction is **exothermic** then an **increase in temperature** will shift the equilibrium to the **left** (in the direction of endothermic change) and the yield will **decrease** i.e. when the temperature is increased by **heat flowing into** the equilibrium mixture, the equilibrium moves in the direction which absorbs heat and lowers the temperature. Note that **increasing** the temperature of the Haber process **decreases** the yield. (see Fig. 13.3.)

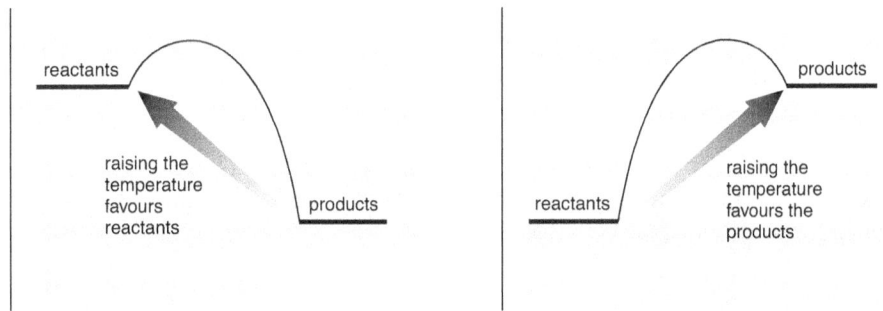

Fig. 13.3

If a reversible reaction is **endothermic** then an **increase in temperature** will shift the equilibrium to the **right** (in the direction of endothermic change) and the yield will **increase** i.e. when temperature is increased, the equilibrium again responds by absorbing heat.

Note if the temperature increases the equilibrium position may shift but, at the same time, both forward and backward rates increase.

Example: In the roasting of limestone to make lime (used in mortar and for neutralising acid soils), the reaction

$CaCO_3(s) \rightleftharpoons CaO(s) + CO_2(g) \quad \Delta H^\ominus = +179 \text{ kJ mol}^{-1}$

is endothermic, so an increase in temperature shifts the equilibrium to the right.

Unit 13 — EQUILIBRIUM

Changing the **pressure** of some gas reactions can alter the **equilibrium position**. Example: In the forward reaction of the equilibrium

$N_2(g) + 3H_2(g) \rightleftharpoons 2NH_3(g)$

there is a decrease in the **number of molecules** (4 gas molecules react to make 2 gas molecules). An increase in pressure causes the equilibrium to shift to the right which decreases the total number of molecules in the mixture, which causes the pressure to decrease.

However, in the equilibrium reaction:

$H_2(g) + I_2(g) \rightleftharpoons 2HI(g)$

pressure does not cause the total number of molecules to change because there are the same numbers of gas molecules (two) on each side of the equation. Changing the pressure has no effect on the position of this equilibrium.

- A **catalyst** has no effect on the equilibrium position and so **no effect** on the yield. Catalysts cause the equilibrium position to be reached **more quickly**. They catalyse the rates of the forward and backward reactions to the **same extents**.

> Remember:
> **Cat**alysts have no effect on Le-**Chat**elier.

- As you have seen above, high pressure and low temperature will increase the yield of NH_3 in the **Haber process**. However low temperature causes the reaction rate to be too slow, so a **compromise temperature** must be used (~**500 °C**). Pressures from **200 to 1000 atm.** are used, higher pressures being uneconomic.

- **Example:** Ethanol may become an important fuel for transport. The ethanol could be made by reacting ethene with steam:

  $CH_2CH_2(g) + H_2O(g) \rightleftharpoons CH_3CH_2OH(g) \quad \Delta H = -287 \text{ kJ mol}^{-1}$

  Consider the above reaction at equilibrium. How would changing the temperature, pressure, and adding a catalyst encourage the yield of ethanol?

  As the forward reaction is exothermic, an **increased temperature** would increase the rate, but the yield would decrease, so a compromise moderate temperature would be required.

  As there are fewer gas molecules on the righthand side, an **increased pressure** would increase both the rate and yield.

  Adding a **catalyst** would increase the rate while not changing the yield, though the yield would be obtained more quickly.

  In conclusion, ethanol production is encouraged by using a high pressure, a moderate temperature and a catalyst.

- **Example:** Methanol, may be used as a way of storing the hydrogen fuel.

  $CO(g) + 2H_2(g) \rightleftharpoons CH_3OH(g) \quad \Delta H = -90 \text{ kJ mol}^{-1}$

  Consider the conditions required to store hydrogen as methanol, and those required to release the hydrogen from the methanol.

  To store the hydrogen the equilibrium must shift to the right which would require a decrease in temperature and an increase in pressure. The increase in pressure and the addition of a catalyst would also increase the reaction rate, so the hydrogen would be stored quickly.

  To release the hydrogen for use as a fuel, the equilibrium must shift to the left which would require an increase in temperature and a decrease in pressure. The increase in temperature and the addition of a catalyst would also increase the reaction rate and so ensure that the hydrogen was stored quickly.

TESTS
RECALL TEST

1 What is meant by 'dynamic equilibrium'?

_____
_____
_____ (1)

2 State Le Chatelier's principle.

_____
_____
_____ (1)

3 This is a general equation for the reaction of an alcohol with a carboxylic acid:

alcohol + acid ⇌ ester + water (this reaction has $\Delta H \sim$ zero)

State which way the position of equilibrium will shift if the following conditions are applied to the reaction (state whether it shifts to the left, right, or does not change):

a [alcohol] is increased _____

b [ester] is increased _____

c [acid] is decreased _____

d temperature is increased _____

e a catalyst is added _____ (5)

4 $CaCO_3(s) \rightleftharpoons CaO(s) + CO_2(g)$ is an endothermic reaction, used to make basic calcium oxide, used to neutralise acid soils. State which way the equilibrium will shift if the following conditions are applied to the reaction (state whether the reaction equilibrium shifts to the left, right, or does not change):

a pressure is increased _____

b temperature is decreased _____

c pressure is decreased _____

d more calcium carbonate is added _____ (4)

Unit 13 — TESTS

5 $H_2(g) + I_2(g) \rightleftharpoons 2HI(g)$ is an exothermic reaction.

State whether the forward rate, backward rate, or yield increase under the following conditions:

a increased temperature _____

b increased pressure _____

c addition of a catalyst _____ (6)

6 Give the stoichiometric (chemical) equation and state the conditions required for these industrial processes:

a production of ammonia _____

b production of sulfur trioxide _____

c oxidation of ammonia _____ (3)

(Total 20 marks)

CONCEPT TEST

1 Consider the equation for the production of ammonia:

$$N_2(g) + 3H_2(g) \rightleftharpoons 2NH_3(g) \quad \Delta H = -92 \text{ kJ mol}^{-1}$$

One industrial design uses a temperature of 450 °C and a pressure of 300-atmospheres with an iron catalyst.

a Explain why a higher temperature is not used, even though the rate would increase.

_____
_____
_____
_____ (2)

b Why is a higher pressure not used?

_____
_____
_____ (2)

c State and explain the effect the catalyst has on the production rate of ammonia.

_____
_____
_____
_____ (2)

d State and explain the effect adding the catalyst has on the yield at equilibrium.

_____

_____
(2)

2 This is the important equilibrium equation for the oxidation of ammonia in a nitric acid plant:

$$4NH_3(g) + 5O_2(g) \rightleftharpoons 4NO(g) + 6H_2O(g) \quad \Delta H_r = -905.6 \, kJ\,mol^{-1}$$

a What is meant by 'dynamic equilibrium' in the context of this reaction?

_____

_____
(2)

b State and explain the effect of increasing the temperature on the position of the reaction equilibrium.

_____

_____
(2)

c Would an increase in pressure increase the yield? Explain your answer.

_____

_____
(2)

d Why is low pressure used, rather than high pressure?

_____

_____
(2)

e The NO made is used to make nitric acid. Write an equation for the formation of nitric acid.

_____

_____
(2)

f Nitric acid may be used with sulfuric acid. State the conditions required to make sulfur trioxide from sulfur dioxide and oxygen.

_____

_____
(2)

(Total 20 marks)

Unit 14: INDUSTRIAL PROCESSES, CATALYSTS, AND THE ENVIRONMENT

- To make chemical production on an industrial scale **economically viable**, chemists balance the **reaction kinetics**, **equilibrium**, and **enthalpy** against **economic** and **environmental** factors.

- **Metal extraction** is important as only some of the metal used in manufacturing is from recycled sources. Most metals are found as ores which are rocks that contain economic amounts of metal. The method used to extract a particular metal depend on its reactivity. When the metal is less reactive than hydrogen then cheap hydrogen is used. When less reactive than carbon then cheap carbon, or more usually carbon monoxide, is used. When more reactive than carbon then electrolysis is required which uses expensive electricity.

  Most metals ores are **oxides** or **sulfides** of the metal. The sulfides are usually roasted in air to form the metal oxide, producing sulfur dioxide. The sulfur dioxide is used to make sulfuric acid. If the gas escaped it would cause acid rain.

- **Tungsten** is found as many ores which are converted into WO_3. The oxide powder is packed into metals tubes and hydrogen is passed over the heated solid:

  $WO_3(s) + 3H_2(g) \rightarrow W(s) + 3H_2O(g)$

  Carbon is not used because the tungsten reacts with carbon to make tungsten carbide, WC. As hydrogen is explosive when mixed with air, care is taken to isolate the hot hydrogen from the air.

- In the **blast furnace**, metal oxides are reduced to the metal using coke, a form of carbon, which is made by baking coal. The metal must be less reactive than carbon. The air blasting into the furnace combusts some of the coke to produce heat:

  $C(s) + O_2(g) \rightarrow CO_2(g)$

  The carbon dioxide produced reacts with more carbon to produce the reducing agent, carbon monoxide:

  $C(s) + CO_2(g) \rightarrow 2CO(g)$

  Usually the carbon monoxide reduces the metal oxide, at a high temperature, to form the metal and carbon dioxide.

- In the case of **iron**, the iron ore haematite is reduced:

  $Fe_2O_3(s) + 3CO(g) \rightarrow 2Fe(l) + 3CO_2(g)$

- In the case of **copper**, copper(II)oxide is reduced:

  $CuO(s) + CO(g) \rightarrow Cu(l) + CO_2(g)$

  The modern method of extracting copper is the direct reduction of the ore chalcopyrite:

  $4CuFeS_2(s) + 11\,O_2(g) \rightarrow 4Cu(l) + 2Fe_2O_3(s) + 8SO_2(g).$

- In the case of **manganese** the main ore is pyrolusite, manganese(IV)oxide, MnO_2, which is first reduced by carbon monoxide:

  $MnO_2(s) + CO(g) \rightarrow MnO(s) + CO_2(g),$

  and then further reduced by carbon:

  $MnO(s) + C(g) \rightarrow Mn(l) + CO(g).$

- **Aluminium** is more reactive than carbon so electrolysis is used. The impure ore, bauxite, is first purified then dissolved in the electrolyte, molten cryolite, which has a much lower melting point then the oxide. The electrodes used are graphite. Electricity is usually made using cheap hydroelectric power. As the process does not use fossil fuels so the process does not produce large quantities of carbon dioxide. At the negative electrode, the cathode:

  $Al^{3+}(l) + 3e^- \rightarrow Al(l)$

  At the negative electrode oxygen is produced:

  $2O^{2-}(l) \rightarrow O_2(g) + 4e^-$

  Recycled aluminium only uses about one tenth of the amount of energy required to make new aluminium by electrolysis, so it is important to recycle aluminium as much as possible.

  > The **cathode** is where reduction occurs.

- **Titanium** cannot be extracted in the blast furnace because it has a very high melting point of 1676°C. The titanium ore, rutile, TiO_2, is first converted into titanium(IV) chloride using chlorine and carbon:

  $TiO_2(s) + 2Cl_2(g) + 2C(s) \rightarrow TiCl_4(g) + 2CO(g)$;

  then the titanium(IV) chloride is reduced using molten sodium:

  $TiCl_4(g) + 4Na(l) \rightarrow Ti(s) + 4NaCl(l)$

  or molten magnesium:

  $TiCl_4(g) + 2Mg(l) \rightarrow Ti(s) + 2MgCl_2(l)$

  Both sodium and magnesium are expensive because they are produced by electrolysis which requires expensive electricity.

- **Recycling scrap** metals are increasingly being used to make metals. The advantage is that less energy is usually required than using any of the above processes and that less ore is dug out of the ground. The disadvantage is that the scrap contains impurities, and the scrap must be sorted and transported to the recycling plant, all of which costs time and money.

  Copper is extracted from low grade ores from the ground water. Scrap iron is added to the water and the iron displaces the copper ions to make copper metal:

  $Fe(s) + Cu^{2+}(aq) \rightarrow Cu(s) + Fe^{2+}(aq)$

- Catalytic **poisons** bind **irreversibly** (permanently) with catalysts and stop them working. Example: **Haem** is a porphyrin complex similar in structure to porphyrin synthetic dyes. This stores oxygen in red blood cells. It is poisoned by CO. (See Fig. 14.1.)

  > Fermentation uses renewable resources but the product is more expensive and impure.

- Biological catalysts called **enzymes** are increasingly used to make useful chemicals. Example: Ethanol is made by **fermentation** of warm sugar solution, using enzyme-containing yeast. Industrially, ethanol is made by **hydrating ethene** over a catalyst of phosphoric acid at 300°C and 70 atm.

- Transition metals have **catalytic properties** due to their **variable oxidation states**. Example: In the Contact Process, V_2O_5 catalyses the overall reaction

  $2SO_2(g) + O_2(g) \rightleftharpoons 2SO_3(g)$

  The reaction happens in two stages:
  (i) $SO_2 + V_2O_5 \rightarrow SO_3 + 2VO_2$
  (ii) $2VO_2 + \frac{1}{2}O_2 \rightarrow V_2O_5$

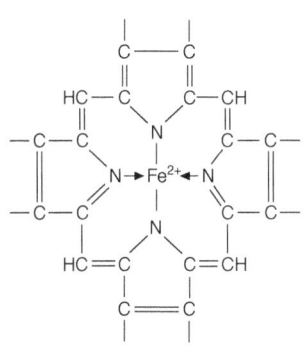

Fig. 14.1 The central part of haem

Unit 14: INDUSTRIAL PROCESSES, CATALYSTS, AND THE ENVIRONMENT

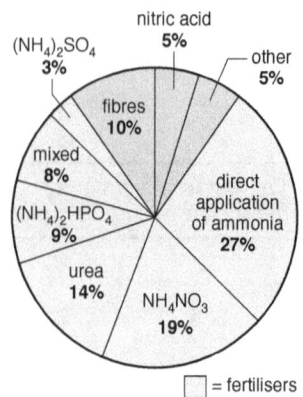

Fig. 14.2 Ammonia uses

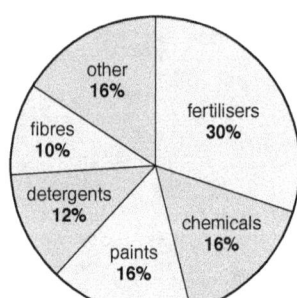

Fig. 14.3 Sulfuric acid uses

> **Explosives** are made by the **nitration** of organic compounds using concentrated nitric acid. Other nitrated compounds are converted into **polyamides** (e.g. Nylon).

- Enormous quantities of ammonia, sulfuric acid, and nitric acid are used industrially to make fertilisers, explosives, and polyamides (see Fig. 14.2). The base ammonia reacts with sulfuric and nitric acids to make the **fertilisers** ammonium sulfate and ammonium nitrate.

  The production of ammonia, the Haber process, was discuscused in full in unit 13.

- Sulfuric acid is made from sulfur trioxide which is manufactured by the **Contact process**. Sulfur dioxide and oxygen react to form sulfur trioxide.
  $$SO_2(g) + \tfrac{1}{2}O_2(g) \rightleftharpoons SO_3(g) \quad \Delta H^\ominus = -197 \text{ kJ mol}^{-1}$$

- The table shows the effects of conditions on rate and SO_3 yield.

	Increase pressure	Increase temperature	Catalyst
Yield	Increases	Decreases	No effect
Rate	Increases	Increases	Increases

  The table shows that the use of high pressure, a compromise temperature, and a catalyst will produce the greatest yield. In practice only **1–2 atm. pressure** is required, with a temperature of **450 °C** and a vanadium(V) oxide **V_2O_5 catalyst**. SO_3 and water react together to form sulfuric acid. However, SO_3 forms a stable mist with water rather than dissolving in it. Instead, SO_3 is **dissolved** in **pure H_2SO_4** which is then **diluted**.

  $SO_3(g) + H_2SO_4(l) \rightarrow H_2S_2O_7(l)$ (oleum)

  $H_2S_2O_7(l) + H_2O(l) \rightarrow 2H_2SO_4(l)$

  A huge amount of sulfuric acid is used in industry (see Fig. 14.3).

- **Nitric acid HNO_3** is made from **ammonia** (see Fig. 14.2). The first step is to pass a mixture of NH_3 and air (O_2) over a platinum/rhodium catalyst at 850 °C to make **nitrogen monoxide** (the reaction is exothermic).

  $4NH_3(g) + 5O_2(g) \rightleftharpoons 4NO(g) + 6H_2O(g)$ (yield approx. 96%)

  Pressure, temperature, and catalysts have the same influence on this reaction as in the Contact process. In practice, only sufficient pressure to pump the gases through the catalyst is required. Only initial (electrical) heating of the catalyst gauze is required as the reaction is exothermic. The NO is further oxidised to **nitrogen dioxide** on contact with more air.

  $2NO(g) + O_2(g) \rightarrow 2NO_2(g)$

  Nitric acid forms when the NO_2 (mixed with more air) dissolves in **water**.

  $4NO_2(g) + O_2(g) + 2H_2O(l) \rightarrow 4HNO_3(aq)$

- Explosives are made by the nitration of organic compounds using concentrated nitric acid. Other nitrated compounds are converted into polyamides (e.g. Nylon).

- Substance **stability** must be discussed in the **context** of a given reaction and with respect to both **thermodynamic** and **kinetic** stability. For example, paper is **thermodynamically unstable** in air with respect to its combustion products; it could oxidise. However, it is **kinetically stable** because the high activation energy means that the oxidation is extremely slow at room temperature.

TESTS
RECALL TEST

1. Which five factors decide whether reaction conditions are economically viable?

   _____
   _____
   _____
   _____
   _____ (5)

2. State five chemical factors that influence the rate of reaction.

   _____
   _____
   _____
   _____
   _____ (5)

3. Which four factors influence the yield?

   _____
   _____
   _____
   _____ (4)

4. If a reaction is exothermic, does high or low temperature increase the yield?
   _____ (1)

5. How do economic factors influence the choice of temperature and pressure?
   _____
   _____ (1)

6. Give the equation for the production of sulfur trioxide from sulfur dioxide and oxygen.
   _____
   _____ (1)

7. For this reaction, state the effect of decreasing pressure on the yield and the rate.
   _____
   _____ (1)

8. Also, state the effect of decreasing temperature on the yield and the rate.
   _____
   _____ (1)

9. Also, state the effect of adding a catalyst on the yield and the rate.
   _____
   _____ (1)

10. State the actual conditions used to produce sulfur trioxide.
    _____
    _____ (2)

Unit 14 — TESTS

11 Sulfur trioxide is converted into sulfuric acid by what?

_____ (1)

12 Give the equation for the oxidation of ammonia to produce nitrogen monoxide.

_____ (1)

13 Give the equation for the oxidation of nitrogen monoxide.

_____ (1)

14 Ammonia, sulfuric acid, and nitric acid are required in large quantities to make what?

_____ (3)

15 Give an example of a nitrogen-containing fertiliser.

_____ (1)

16 Why were lead compounds added to petrol in the past?

_____ (1)

17 Give two different reasons for the phasing out of leaded fuels.

_____ (2)

18 What is the function of catalytic converters?

_____ (3)

19 What metal is used in catalytic converters?

_____ (1)

20 How do poisons stop catalysts working?

_____ (1)

21 How is ethanol made industrially?

_____ (1)

22 Give a use for ethanol.

_____ (1)

23 Give a different use for methanol.

_____ (1)

(Total 40 marks)

CONCEPT TEST

1. With crude oil prices increasing, methanol production could be developed to replace petrol, or methanol could be used to lessen pre-ignition in petrol.

   One way of producing methanol is from the exothermic reaction of carbon monoxide and hydrogen, which are produced from coal and water. This methanol reaction is of industrial importance:

   $CO(g) + 2H_2(g) \rightarrow CH_3OH(g)$ using a $ZnO/CrO_3(s)$ catalyst.

   a What is pre-ignition, and why is it costly?

   _____
   _____
   _____ (2)

   b Why is a catalyst used to produce methanol?

   _____
   _____ (2)

   c What would be the effect of increasing the pressure on the reaction?

   _____
   _____
   _____ (2)

   d How would increasing the temperature influence the reaction?

   _____
   _____
   _____ (2)

   e Why would widespread use of methanol as a fuel improve the environment?

   _____
   _____
   _____ (2)

   f How do catalytic converters improve the environment?

   _____
   _____
   _____ (3)

   g Is the catalyst homogeneous or heterogeneous? Give your reasons.

   _____
   _____
   _____ (2)

   (Total 15 marks)

Unit 15 — EXPERIMENTAL SKILLS

- There is not enough space here to give detailed descriptions of the methods used in practical chemistry. The intention of these notes is to show you how to use or describe **practical techniques** in written exams or assessed practicals.

To ensure all reactant molecules have **full contact** with each other, mixtures of liquids must be **mixed** by stirring or shaking.

The term **heat strongly** is associated with the thermal decomposition of compounds. You should **heat gently** when trying to speed up a reaction without boiling away solvent water or causing decomposition. Use an **oil bath** to heat above 100 °C. A bare flame causes localised hot-spots. An **electric heater** is especially useful when heating flammable liquids.

The **melting point** of a compound (up to about 250 °C with an oil bath) is determined by using a capillary tube and thermometer (see Fig. 15.1).

The **boiling point** is measured by placing a thermometer **just above** the surface of the boiling liquid. A flask is a convenient container (see Fig. 15.2).

Heating or boiling **under reflux** speeds up a reaction without losing volatile (readily vaporised) reactants or products. A flask with a vertical condenser attached is used (see Fig. 15.3).

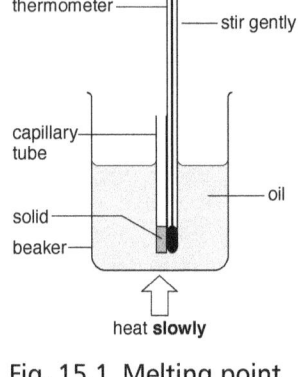

Fig. 15.1 Melting point apparatus

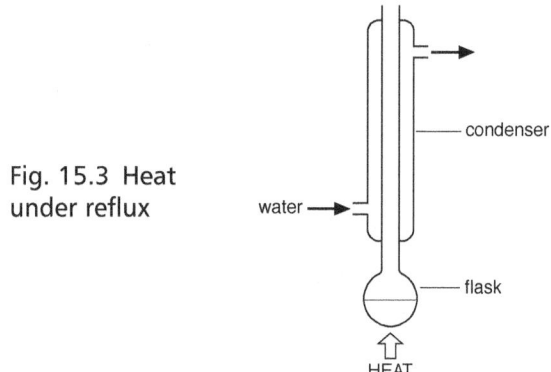

Fig. 15.3 Heat under reflux

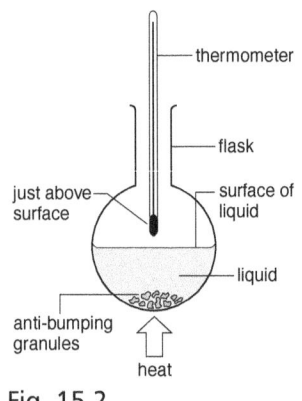

Fig. 15.2

- Use **fractional distillation** to separate two volatile liquids. The liquid mixture is gently boiled. The vapour contains a **higher** proportion of the **more volatile** substance (the one with the **lower** boiling point). NB it is not correct to say that the more volatile substance boils first. The vapour mixture separates as it rises inside the fractionating column. The **less volatile** substance condenses and falls back as a **liquid** into the flask. The **more volatile** substance reaches the **top** of the column as a **vapour** (see Fig. 15.4).

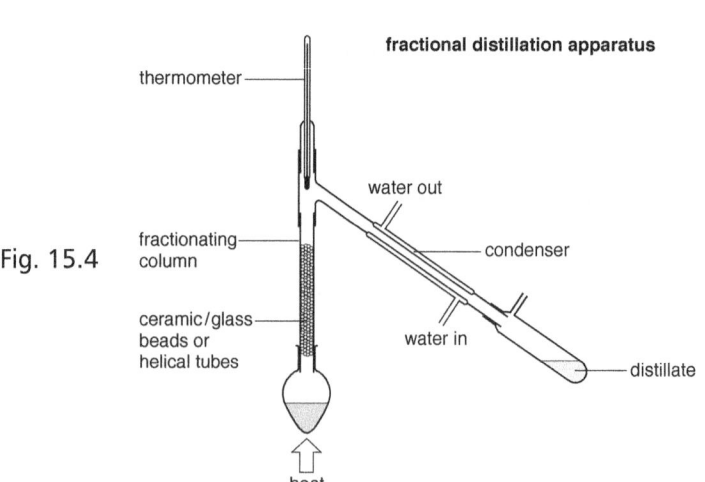

Fig. 15.4

- Use **filtration** to separate a solid from a liquid. The two key terms associated with filtration are **filtrate** (the liquid that passes through the filter paper) and **residue** (the solid trapped by the filter paper).

  You **filter under reduced pressure** to quickly separate a solid from a liquid or to filter a hot mixture before it cools. The apparatus is a **Buchner funnel** and (side-arm) **filter flask** attached to a vacuum pump (driven by the flow from a tap) (see Fig. 15.5).

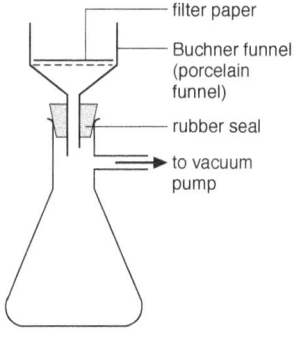

Fig. 15.5

- Details of **safety points** must be included in assessed practicals and they are often examined in written papers. The examiners assume that safety glasses and lab coats are worn, but **mention them anyway**. Always give a reason for your safety point.

  Watch out for **flammable** substances and include the phrase use **no naked flames** in your practical instructions.

  Where poisonous solids are used add the obvious phrase '**do not eat**' and include the use of protective gloves.

  Volatile and gaseous **poisons** must be used in a **fume cupboard**.

- **Precision** is important. Many instruments can usually provide highly accurate data, but if you are asked to state a measurement to only 1-°C or to 2 significant figures **then do so**. The **least accurate** measurements (those with the least number of significant figures) **limit** the overall experimental accuracy. If data is correct to **2** significant figures, you **cannot** give the result of a calculation to **3** significant figures.

  Remember when recording results to always draw a table with a box around it., record titration results to two decimal places (the second figure after the decimal place must be a 0 or 5, e.g. 25.15 or 25.20).

- Always add **units** to your results when when showing your calculations in calculations. Always show some steps in your calculations as they earn marks.

- The **tests** for inorganic **cations** and the **flame** test colours may occur in your assessment. The addition of **NaOH(aq)** to aqueous cations is very important, so check you know the reaction of $OH^-(aq)$ with:
  Mg^{2+}, Ca^{2+}, Sr^{2+}, Ba^{2+}, Al^{3+}, Zn^{2+} and **transition metal** cations.
  Also remember that heating any **ammonium** NH_4^+ salt with NaOH(aq) produces **ammonia** NH_3 gas, e.g.
  $NH_4^+(aq) + OH^-(aq) \rightarrow NH_3(g) + H_2O(l)$

Flame colours
Li red (crimson) Na yellow Ca orange-red Sr red (crimson) Ba pale green (apple green) K lilac or pale purple* Mg compounds **do not** produce a coloured flame
*You should state that the lilac colour is visible through blue glass.

- Here are the **tests** for inorganic **anions**.
  Halide ions form precipitates with $AgNO_3(aq)$ (see unit 4 for details);
  Carbonates CO_3^{2-} produce CO_2 when added to acid (fizzing – effervescence) and when heated strongly (except group-1 carbonates).
  Hydrogencarbonates HCO_3^- (only group 1 stable as solids) such as $NaHCO_3$ also produce CO_2 when added to acid and when gently heated.
  Sulfites SO_3^{2-} produce SO_2 with acid, and form a precipitate with $Ba^{2+}(aq)$ that dissolves in acid.
  Sulfates SO_4^{2-} form a heavy white precipitate with $Ba^{2+}(aq)$ that is insoluble in acid.
  Hydrogensulfates HSO_4^- (as in $NaHSO_4$) react like sulfate ions. Their aqueous solutions are very acidic.

Unit 15 — EXPERIMENTAL SKILLS

PLANNING PRACTICALS

It is quite easy to become very confused and stressed when **planning** practical work for assessment or **describing** a practical in detail during a written exam. Do the planning in the following order to establish a **framework**:

(A) Read the **instructions** given.

(B) Outline the **chemistry** involved.

(C) Outline a **method** (without amounts or descriptions of apparatus).

(D) Read the instructions again (have you **missed** something?).

(E) Calculate suitable **amounts** in moles of the reactants.

(F) Now choose your **apparatus**.

(G) Remember to **stir** or **shake** mixtures. State **when** to make **measurements** such as time or temperature (start and finish). Are you **heating**: if so, how strongly? **Repeat** the experiment keeping all independent variables constant and calculate an **average** for each set of measurements.

(H) **Safety** points.

(I) **Summarise** the practical procedure, give it the title **Practical Outline** and put it at the beginning of your written account.

(J) State the **assumptions** you are making (about e.g. specific heat capacity, heat loss, purity of chemical reagents).

(K) **Accuracy** must usually be assessed. State all sources of error and indicate which are the main ones. Assumptions are often sources of errors.

(L) **Re-read** the instructions, including those that seem trivial or obvious.

> Stir the liquids as soon as they are poured together.

> A good place to start is 2-cm^3 of liquid reactants or 1–2 g of solid reactant. Aim for solution concentrations from 0.01 to 0.2-mol dm^{-3}.

> Remember also that solutions must be made up accurately.

- **Accuracy** improves if the experiment is repeated and an **average** calculated, so always say that the experiemnet must be repeated. Accuracy is a reflection of how accurate the apparatus is. It may be the results are accurate, but not reliable.

  Remember also that solutions must be made up accurately. This means graduated pipettes and volumetric flasks must be used.

  Reliable results are those that may be trusted. Usually the values for reliable results are close together or will produce a smooth curve on a line graph.

  Example :

  These results are accurate but not reliable: 25.30, 26.85, 10.26.

  These results are relable but not as accutate: 25.0, 25.1, 25.1.

  These results are both accurate and relaible: 25.15, 25.15, 25.20.

  Anomolous result are those that are far away from the rest of the results. They could be far away from a smooth curve, or far from a cluster of results are said to be. In your written analysis, point out any anomolous results.

- **Rate** experiments usually involve measuring **time**. Remember to state clearly the start time and the stop time. Any of the five main factors (concentration, pressure, temperature, surface area, and catalyst) may influence the rate, so make sure you have allowed for the relevant ones. Light may also have an influence.

- The **enthalpy change** for a reaction taking place in solution can be calculated from **temperature changes** and measurements of **mass/volume** (grams/cm^3 of water) and **amount** (moles of reactants). You must remember to state clearly the temperatures you read **before the start** of the reaction and **at the end**. Remember also that **inaccuracies** are introduced by **heat loss** to the container or to the air (by conduction or evaporation). **Incomplete reaction** may be caused by insufficient stirring (see unit 11).

TESTS
RECALL TEST

1. Draw a simplified graph to show how fractional distillation occurs in steps within the apparatus.

(2)

2. State how ammonium ions may be detected in solution.

___
___
___

(3)

3. Which substances/ions are indicated by these observations?

   a It forms a cream ppt with $AgNO_3$(aq)

   ___
   ___

   b It effervesces with HCl(aq), and limewater turns milky

   ___
   ___

   c It forms a white ppt with Ba^{2+}(aq), which dissolves in HCl(aq), and produces a gas which turns orange dichromate ion green

   ___
   ___

   d It forms a white ppt with Ba^{2+}(aq), which does not dissolve in HCl(aq)

   ___
   ___

   e A solid heated strongly produces a brown gas and a gas that relights a glowing splint

   ___
   ___

(5)

(Total 10 marks)

Unit 15 TESTS

CONCEPT TEST

1. Aqueous sodium hydrogen sulfite may be used to detect a carbonyl compound in analysis. Aldehydes and ketones will form a white precipitate with $NaHSO_3(aq)$. Once the white solid formed is purified, its melting point is determined. The carbonyl compound originally present may be ascertained by reference to a data book which lists the melting points of sodium hydrogen sulfite derivatives.

   a Explain how the white solid may be recrystallised.

   _____
   _____ (4)

   b Explain how to determine the melting point of the purified solid.

   _____
   _____
   _____ (3)

   c The boiling point would help to confirm which carbonyl compound was present. Describe how you would determine the boiling point.

   _____
   _____
   _____ (2)

2. White solid X dissolves in water to form a colourless solution. A sample of this solution when mixed with aqueous sodium sulfate forms a heavy white precipitate Y. When solid X is heated strongly, a brown gas Z evolves and a glowing splint readily relights. The white solid produces a pale green flame colour.

   White solid P dissolves in water. The solution when mixed with hydrochloric acid produces a colourless gas Q that will turn aqueous acidified potassium dichromate from orange to green. When the solid P is put into a flame a pale lilac colour is produced, which is visible through blue glass.

   a State the formula of:

   X _____

   Y _____

   Z _____ (6)

   b State the formula of:

   P _____

   Q _____ (2)

   c State how you would confirm the presence of bromide ions in a solution.

   _____ (3)

   (Total 20 marks)

Unit 16 — CHEMICAL CALCULATIONS 1

The answers to the **Exercises** are at the end of this section after the tests.

- **Calculating atom economy** is becomeing increasingly important because it is a measure of how much waste is produced by a chemical process.

$$\text{Percentage atom economy} = \frac{\text{molecular mass of the useful products} \times 100}{\text{total molecular mass of the reactants}}$$

If the % atom economy was 100% that would mean there was no waste. however if the % atom economy was 10% then only 10% was useful product, and the rest was waste which would be expensive to dispose of, and may have an evironmental impact.

Examples: One way of extracting copper is:

$4CuFeS_2(s) + 11\ O_2(g) \rightarrow 4Cu(l) + 2Fe_2O_3(s) + 8SO_2(g)$.

Percentage atom economy = 254 x 100 / 1086 = 23.4%

Contrast this poor atom economy with the production of methanol from hydrogen and carbon monoxide:

$CO(g) + 2H_2(g) \rightarrow CH_3OH(g)$

Percentage atom economy = 32 x 100 / 32 = 100%

One mole of a substance contains the same number of particles as one mole of any other substance.

- The **amount of substance** (symbol n) is measured in moles. **One mole** (1-mol) is the amount of a substance that contains the same number of particles as there are atoms in exactly 12 g of carbon-12. The examiner may require you to recall this definition.

Examples: 1 mol of water H_2O contains 1 mol of water molecules, 2 mol of hydrogen atoms, and 1 mol of oxygen atoms.

1 mol of magnesium chloride contains 1 mol of Mg^{2+} and 2 mol of Cl^- ions.

The mass of 1 mol of water is 18.015 28 g i.e. [(2 × 1.007 94) + 15.9994] g.

Exercise 1: How many moles of atoms in the following?

a 1 mole of hydrogen, H_2;   b 2 moles of methanse CH_4;
c 0.5 moles of NaOH;   d 1 mole of H_2SO_4.

- The **mass** of one mole of atoms of an **element** (e.g. Ne, Fe, Cl, but not Cl_2) is equal to the relative atomic mass RAM in grams. (There is a periodic table at the back of the book).

Example: The mass of one mole of Cl **atoms** is 35.5 g.

Exercise 2: Determine the mass in each case:

a 1 mole of Na;   b 4 moles Cl;
c 2 moles of O;   d 0.25 moles of sulfur.

- The **mass** of one mole of a **molecular compound** is equal to the formula mass (i.e. the sum of the RAM values) in grams.

Example: The mass of one mole of Cl_2 molecules is 35.5 × 2 = 71.0 g.

Ionic compounds consist of separate ions. It is not correct to assign a relative molecular mass to these substances. You should refer to the **relative formula mass M_r** (i.e. the sum of the RAM values) in grams.

Example: The mass of 1 mol of magnesium chloride is 95.2104 g i.e. [24.3050 + (2 × 35.4527)] g.

The **simplest** approach is to refer to all substances in terms of their **formula masses**. The formula mass is the mass of 1 mol.

1 mol of $MgCl_2$ (95.2104-g) contains 6.023×10^{23} Mg^{2+} ions and 2-×-6.023-×-10^{23} Cl^- ions.

Example: The formula mass of magnesium chloride $MgCl_2$ is 95.2104 g.

Exercise 3: Calculate the amount in moles in each case:
a   12 grams of C;
b   40 grams of NaOH;
c   8 gram of calcium;
d   46 grams of Na.

Exercise 4: Calculate the formula mass of the substance in each case:
a   2 moles has a mass of 60 grams;
b   64 grams is one mole;
c   0.5 moles has a mass of 23 grams;
d   0.1 moles has a mass of 4 grams.

Exercise 5: Calculate the mass in each case:
a   2 moles of CO_2;
b   0.1 moles of water;
c   3 moles NaOH;
d   0.5 moles of CaO.

Exercise 6: Calculate the moles in each case:
a   98 grams of sulfuric acid. H_2SO_4;
b   9 grams of water;
c   30 grams of NaOH;
d   20 grams of calcium carbonate, $CaCO_3$.

Exercise 7: Calculate the relative formula mass in each case:
a   2 moles of a nitrogen oxide has a mass of 60 grams;
b   0.5 moles of a sulfur oxide has a mass of 32 grams;
c   1 mole of a nitrogen oxide has a mass of 46 grams;
d   0.1 moles of a lead oxide has a mass of 22.3 grams.

> When **stuck** for ideas, try converting the **given quantities** (masses, volumes/concentrations) into **amounts** in moles. Most of the mathematical expressions work via moles. Usually one calculation links to another via the **reacting ratios** shown by the balanced chemical equation.

- You may find you can work out simple cases in your head, but it is safer in the long run to remember an equation.

$$\textbf{amount (moles)} = \frac{\textbf{mass (grams)}}{\textbf{formula mass (grams)}} \quad \text{i.e. MOLES} = \frac{\text{mass in grams}}{\text{mass of one mole}}$$

If desperate,

$$\text{moles} = \frac{\text{GRAMs}}{\text{RAMs}}$$

where RAM is relative atomic mass(es).

- The **number** of atoms in 1 mol of carbon-12 (12 g exactly) is 6.023×10^{23} and is called the **Avogadro constant** (symbol L).

- You must know how to **calculate** the **number of particles** (atoms, molecules, or ions, etc.) in a given mass of a substance. One mole of a substance (equivalent to its formula mass) contains L particles.

Examples: 1 mol of Ne (20.1797 g) contains 6.023×10^{23} Ne atoms.
1 mol of H_2O (18.015 28 g) contains 6.023×10^{23} H_2O molecules, 2–× 6.023×10^{23} H atoms, and 6.023×10^{23} O atoms.
1 mol of $MgCl_2$ (95.2104 g) contains 6.023×10^{23} Mg^{2+} ions
and $2 \times 6.023 \times 10^{23}$ Cl^- ions.

Exercise 8: Calculate the number of atoms in each case:
a   2 moles of NaOH;
b   0.01 moles of $CaCO_3$;

- Occasionally, you will need to derive a mass or a volume from a given **density** value. The SI unit of density is kg m^{-3}, which is equivalent to the more practical unit of g cm^{-3}.

$$\textbf{density (g cm}^{-3}\textbf{)} = \frac{\textbf{mass (g)}}{\textbf{volume (cm}^3\textbf{)}} \quad \text{i.e. mass = density} \times \text{volume}$$

Unit 16 CHEMICAL CALCULATIONS 1

A **structured** approach to **problem solving** is to **first** extract the data and **then** look for combinations of data that allow you to use an appropriate equation.

- When required to **calculate masses** using chemical equations the mole concept should be used. A chemical equation shows the mole ratio of the reactants and products, in that the molecular ratio is the same as the mole ratio.

  Example: nitrogen monoxide reacts with oxygen to make nitrogen dioxide:

  $2NO(g) + O_2(g) \rightarrow 2NO_2(g)$

  Here 2 moles of $NO(g)$ react with one mole of $O_2(g)$ to make 2 moles of $NO_2(g)$. How many grams of oxygen will react if 120 gram of $NO(g)$ is present?

  (i) Convert the mass into moles:

  Moles of NO = mass / formula mass = 120 / 30 = 4

  (ii) Use the mole ratio find the moles of oxygen:

  2 moles of $NO(g)$ react with 1 mole of $O_2(g)$, ratio: 2 : 1,

  so 4 moles of $NO(g)$ react with 2 mole of $O_2(g)$

  (iii) Convert the moles of oxygen into grams:

  mass = moles x formula mass = 2 x 32 = 64 grams.

  So 120 grams of nitrogen monoxide react with 64 grams of oxygen.

  Exercise 9:

  a $Fe_2O_3(s) + 3CO(g) \rightarrow 2Fe(l) + 3CO_2(g)$

  If one tonne of iron(III) oxide is converted into iron,

  (i) What mass of iron, in grams, will be made?

  (ii) How many tonnes of carbon dioxide would be made?

  b $CH_2CH_2(g) + H_2O(g) \rightleftharpoons CH_3CH_2OH(g)$

  If one kilogram of ethanol was required, how many grams of ethene would be required?

- You will be expected to **convert** a **volume** of gas into an **amount** in moles, or to convert an amount of a gas in moles into its corresponding volume. You will be told that one mole of gas occupies 24 dm^3 at room temperature and pressure (or some other volume for different conditions). This volume is called the **molar gas volume**. As two moles of a gas would occupy twice the volume of one mole, then it is simple to remember that:

  volume of gas = number of moles × volume of one mole of gas

At room temperature the volume of water would be negligible because of condensation.

- The **ratio** of the **volumes** of gases in a reaction is equal to the ratio of the **amount** (number of moles) of each gas in the balanced equation.

  Example: When 100 cm^3 of methane burns in air, the volumes of the gases involved may be calculated from the balanced equation. Note that all volumes are measured at the same temperature and pressure:

  $CH_4(g) + 2O_2(g) \rightarrow CO_2(g) + 2H_2O(g)$
  100 cm^3 200 cm^3 100 cm^3 200 cm^3 (of steam)

  Exercise 10:

  $C_2H_4(g) + 3O_2(g) \rightarrow 2CO_2(g) + 2H_2O(g)$

  (i) If 200 cm^3 ethene reacted with oxygen, what volume of carbon dioxide would be made?

  (ii) How many moles of water would be made?

  (iii) What would be the mass of the water?

- You will often have to calculate the **percentage yield** of a reaction.

Usually an organic compound undergoing reaction is not in excess; in redox reactions, the acid is in excess.

$$\text{percentage yield} = \frac{\text{actual amount (mol)}}{\text{maximum possible amount (mol)}} \times 100\%$$

Sometimes, one of the reactants will be in **excess**. When calculating yield, be sure to first calculate the amounts (in mol) of all the reactants and base the theoretical yield on the reactant that is present in the **smallest amount**.

TESTS
RECALL TEST

1 a Give an equation for moles, using mass and RAM.

_____

_____

_____ (1)

b Give an equation for mass, using moles and RAM.

_____

_____

_____ (1)

c Give an equation for RAM, using moles and mass.

_____

_____

_____ (1)

2 a If one mole of gas occupies 24 dm^3, calculate the gas volume of:

   i   3 mol methane

   ii 0.5 mol oxygen

(2)

b How many moles are there in 1000 dm^3 (1 m^3) of methane?

_____

_____ (1)

3 Give the equation for density.

_____

_____

_____ (1)

(Total 10 marks)

Unit 16 — TESTS

CONCEPT TEST

1 A carbonate of a metal M has the formula M_2CO_3. This is the equation for the reaction of the carbonate with hydrochloric acid:

$$M_2CO_3(s) + 2HCl(aq) \rightarrow 2MCl(aq) + H_2O(l) + CO_2(g)$$

0.49 grams of the solid reacted with the acid to make exactly 84 cm³ of gas.

    a Calculate the number of moles of carbon dioxide produced,

    b Calculate the number of moles of $M_2CO_3(s)$,

    c Calculate the relative formula mass of the carbonate,

    d Calculate the relative atomic mass of the metal M, and so identity it.

(5)

2 Compare the atom economy of these two reactions that both produce ethene:

Reaction 1: $CH_3CH_2Cl + KOH \rightarrow C_2H_4(g) + KCl(aq) + H_2O(l)$

Reaction 2: $CH_3CH_2OH(l) \rightarrow C_2H_4(g) + H_2O(l)$

    a Calculate the atom economies:

        Reaction 1:

        Reaction 2:

    b Based on the idea of atom economy, which is the better reaction ?

(5)

(Total 10 marks)

Answers to the exercises:

Exercise 1:
a 2 moles, **b** 10 moles; **c** 1.5 moles; **d** 7 moles.

Exercise 2:
a 23 grams; **b** 142 grams; **c** 32 grams; **d** 8 grams.

Exercise 3:
a 1 mole; **b** 1 mole; **c** 0.2 moles; **d** 2 moles.

Exercise 4:
a 30; **b** 64; **c** 46; **d** 40.

Exercise 5:
a 88 grams; **b** 1.8 grams; **c** 120 grams; **d** 28 grams.

Exercise 6:
a 1 mole; **b** 0.5 moles; **c** 0.75 moles; **d** 0.2 moles.

Exercise 7:
a 30 (NO); **b** 64 (SO_2); **c** 46 (NO_2); **d** 223 (PbO).

Exercise 8:
a 12.046×10^{23} atoms = 1.2046×10^{24} atoms ;
b 30.115×10^{23} atoms = 3.0115×10^{22} atoms;

Exercise 9:
a (i) One tonne of iron = 1 000 000 grams.
Moles of iron(III) oxide = 1 000 000 / 160 = 6250 moles,
Ratio is 1:2,
6250 : (6250 x2) = 12 500 moles of iron,
mass of iron = 12 500 x 56 = 700 000 grams of iron.
(ii) Moles of iron(III) oxide = 1 000 000 / 160 = 6250 moles,
Ratio is 1:3,
6250 : (6250 x3) = 18 750 moles of carbon dioxide,
Mass of carbon dioxide = 18 750 x 44 = 825 000 grams = 0.825 tonnes.

b $CH_2CH_2(g) + H_2O(g) \rightleftharpoons CH_3CH_2OH(g)$
If one kilogram of ethanol was required, how many grams of ethene would be required?
1 kg = 1000 grams
moles of ethanol = 1000 / 46 = 21.7 moles
Ratio is 1:1,
moles of ethene = 21.7 moles
Mass of ethene = 21.7 x 28 = 609 grams.

Exercise 10:
$C_2H_4(g) + 3O_2(g) \rightarrow 2CO_2(g) + 2H_2O(g)$
(i) 400 cm^3
(ii) Moles of ethene = 200 / 24 000 = 0.0083 moles.
Ratio is 1 : 2
Moles of water = 0.0083 moles x 2 = 0.0167 moles
(iii) Mass = moles x formula mass = 0.0167 x 18 = 3 grams.

Unit 17

CHEMICAL CALCULATIONS 2

> Some students prefer to remember
> mol = conc. × vol.

> Never use the symbol M as an abbreviation for mol. It is an obsolete term which means mol dm^{-3}.

- The **molar concentration** of a solution describes the amount of solute dissolved in a given volume of solution (usually 1 dm^3). You will have to calculate the **molar concentration** of solutions or, if given a concentration, calculate the moles of solute present. The units of concentration are mol dm^{-3} i.e. $\frac{\text{mol}}{\text{dm}^3}$ so it should not be difficult to remember that

$$\text{concentration (mol dm}^{-3}\text{)} = \frac{\text{amount of solute (mol)}}{\text{solution volume (dm}^3\text{)}}$$

Some students prefer to remember: **mol = conc. × vol**.

If desperate,
$$\text{conc.} = \frac{\text{mols}}{\text{vols (in dm}^3\text{)}}$$

Example: When 2 moles of NaOH are dissolved in 4 dm^3 then the concentration is 0.5 mol dm^{-3}.

Exercise 11: Calculate these concentrations;
a 4 mols of NaOH is dissolved in 2 dm^3;
b 0.5 moles of NaCl is dissolved in 2 dm^3;
c 0.5 dm^3 of solution holds 0.01 moles of ammonia;
d 0.250 dm^3 contains 0.05 moles of HCl.

> The answers to the **Exercises** are at the end of this section before the tests.

Exercise 12: What amount of chemical is present? (Hint: amount means moles)
a 2 dm^3 of 3 mol dm^{-3}, NaOH; **b**; 0.5 mol dm^{-3}, volume 2 dm^3;
c a flask that holds 0.500 dm^3 KOH, has a concentrtion of 0.1 mol dm^{-3};
d a bottle of 2 mol dm^{-3} hydrochoric acid has a volume of 0.250 dm^3.

Exercise 13: What volume, in dm^3, is required to hold the stated amount?
a 2 moles is required of a 0.5 mol dm^{-3} NaOH solution;
b a container for a 5 mol dm^{-3} solution is to hold 1 mole of HCl;
c 0.1 moles is required using a 0.05 mol dm^{-3} solution;
d A solution is 0.01 mol dm^{-3}, and 2 moles is required.

- 1 decimetre cubed (1 dm^3) = 1 litre (1 l) = 1000 millilitres (ml)

$$1 \text{ dm}^3 = 1000 \text{ cm}^3 = 1000 \text{ ml} \qquad \text{dm}^3 = \frac{\text{cm}^3}{1000}$$

Use dm^3 and cm^3 for all calculations.

Example: When 25 cm^3 is used in a titration the volume is 0.025 dm^3.

Exercise 14: Convert the volumes in cm^3 into dm^3.
a 20 cm^3; **b** 2000 cm^3; **c** 4 cm^3; **d** 400 cm^3.

Exercise 15: Convert the volumes in dm^3 into cm^3.
a 2 dm^3; **b** 0.25 dm^3; **c** 0.001 dm^3; **d** 24 dm^3.

- A solution of known concentration that is used to determine the concentration of another solution is a called a **standard solution**.

- Often you will be given concentration in grams per decimetre cubed, g dm^{-3}. Just convert the grams into moles using the usual equation:

$$\frac{\text{amount}}{\text{(moles)}} = \frac{\text{mass (grams)}}{\text{formula mass (grams)}}$$

and that will give you the concentration in mol dm^{-3}.

Many students find the **calculation map a** useful way to see how the different equations link (see Fig. 17.1).

Example: In a practical a solution was labelled, 50 g dm^{-3} Na$_2$CO$_3$.12H$_2$O.

Formula mass of Na$_2$CO$_3$.12H$_2$O = (2 × 23) + 12 + (3 × 16) + (12 × 18) = 322.

Moles = mass / formula mass = 50 / 322 = 0.16 moles, so the solution concentration = 0.16 moles dm^{-3}.

Exercise 16: Convert these concentrations into mol dm^{-3};
- **a** NaOH, 8 g dm^{-3};
- **b** 1 g dm^{-3} CaSO$_4$.H$_2$O;
- **c** KMnO$_4$, 3.18 g dm^{-3};
- **d** sea water contains 35 g dm^{-3} NaCl.

- In titration calculations, you are given:

  (i) the concentrations and volumes of solution X in one piece of glassware (burette or pipette),

  and (ii) the volume of solution Y of unknown concentration in the other piece of glassware.

Work in three stages:
(A) Calculate the moles of substance present in solution X above.
(B) Use the balanced equation for the reaction to find the molar ratios of the reacting substances X and Y and hence the amount in moles of substance Y present in (ii).
(C) Use the amount of Y in moles present and the volume of its solution to calculate the concentration. Of course, the examiners set this calculation regularly, so they are likely to add an extra twist, such as including a dilution.

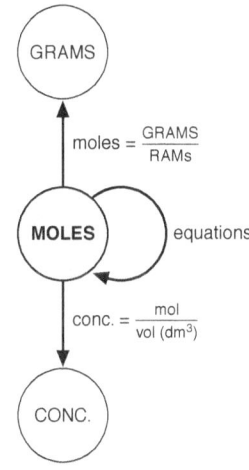

You could add other calculation steps to this diagram.

Fig. 17.1

A sample calculation:
Potassium permanganate concentration 0.020 mol dm^{-3} was used to standardise a solution of iron(II) sulfate. The titration flask held a 25.0 cm^3 **aliquot** (the solution measured by pipette) of iron(II) sulfate and 25.0 cm^3 of sulfuric acid 0.1 mol dm^{-3}. The average **titre** (volume delivered by the burette) was 27.93 cm^3. Write an ionic equation for the reaction between iron(II) sulfate and potassium permanganate.
Calculate the iron(II) sulfate concentration.

The **ionic equation** is
$$MnO_4^-(aq) + 5Fe^{2+}(aq) + 8H^+(aq) \rightarrow Mn^{2+}(aq) + 5Fe^{3+}(aq) + 4H_2O(l)$$

Extract the data:
KMnO$_4$ concentration = 0.02 mol dm^{-3}; vol. = 27.93 cm^3.
FeSO$_4$ concentration unknown; vol. = 25.0 cm^3.

Calculation:
(A) Amount KMnO$_4$ (mol) = $0.02 \times \frac{27.93}{1000}$ = 0.000 5586 mol

Unit 17 — CHEMICAL CALCULATIONS 2

(B) Molar ratios: 1 mol MnO_4^-(aq) reacts with 5 mol Fe^{2+}(aq)
Amount of $FeSO_4$ (mol) = 0.002 793 mol

(C) Concentration of $FeSO_4$ (mol dm^{-3}) = $\dfrac{0.002\,793}{25/1000}$ = 0.111 72 mol dm^{-3}

The data is given to **two significant figures or better**, so the answer must be given to the **same** degree of accuracy i.e. concentration $FeSO_4$ = 0.11 mol dm^{-3}.

Answers to the exercises:

Exercise 11: (it does not matter which chemical is present. Just use the number of moles and volume in dm^3, to calculate the concentration).
a 2 mol dm^3; **b** 0.25 mol dm^3; **c** 0.02 mol dm^3; **d** 0.02 mol dm^3.

Exercise 12:
a 6 moles; **b** 1 moles; **c** 0.05 moles; **d** 0.5 moles^3.

Exercise 13: (Hint: volume in dm^3 = moles / concentration)
a 4 dm^3; **b** 5 dm^3; **c** 2 dm^3; **d** 200 dm^3.

Exercise 14:
a 0.020 dm^3; **b** 2 dm^3; **c** 0.004 dm^3; **d** 0.400 dm^3.

Exercise 15:
a 2000 cm^3; **b** 250 cm^3; **c** 1 cm^3; **d** 24000 cm^3.

Exercise 16:
a 0.2 mol dm^{-3}; **b** 0.0065 mol dm^{-3};
c 0.02 mol dm^{-3}; **d** 0.60 mol dm^{-3}.

TESTS

RECALL TEST

1 a Give an equation for concentration (in mol dm^{-3}).

   _____
   _____ (2)

  b Give an equation for moles using concentration and volume.

   _____
   _____ (2)

  c Give an equation for volume (cm^3) using concentration and moles.

   _____
   _____ (2)

2 a Give the equation that is used to convert concentration in g dm^{-3}. to concentration in mol dm^{-3}.

   _____
   _____ (1)

  b Give the equation that is used to convert concentration in mol dm^{-3}. to concentration in g dm^{-3}.

   _____
   _____ (1)

  c Give an equation for percentage yield.

   _____
   _____
   _____ (2)

CONCEPT TEST

1 During a titration of ethanoic acid against potassium hydroxide, the titre was found to be 23.45-cm^3 ethanoic acid against a 10.00-cm^3 KOH aliquot. If the alkali was 0.60-mol dm^{-3}, what is the concentration of the ethanoic acid?

   $$CH_3COOH(aq) + KOH(aq) \rightarrow CH_3COO^-K^+(aq) + H_2O(l)$$

   _____
   _____
   _____
   _____ (2)

2 Often iodine is titrated against thiosulfate. Write an equation for the reaction. If a titre of 26.85-cm^3 potassium thiosulfate solution, 0.020-mol dm^{-3}, reacted with 50.0-cm^3 iodine, dissolved in potassium iodide, what would be the-iodine concentration?

   $$2K_2S_2O_3(aq) + I_2(aq) \rightarrow K_2S_4O_6(aq) + 2KI(aq)$$
   (3)

Unit 17 — TESTS

i an aqueous solution of 3.16-g·dm⁻³ potassium manganate(VII), KMnO$_4$, labelled solution P,

ii an aqueous solution of ethanedioic (oxalic) acid, H$_2$C$_2$O$_4$, of unknown concentration, labelled Q,

ii an aqueous solution of 7.865-g·dm⁻³ vanadium(III) chloride, VCl$_3$, labelled R,

iii aqueous bench sulfuric acid.

3 Silver chloride is less soluble than silver chromate, so the chloride ion concentration may be determined by titrating aqueous silver nitrate into a chloride ion solution. A few drops of yellow potassium chromate, here being used as an indicator, will produce a red silver chromate precipitate (mixed in with the yellow). If the aliquot was 25.0-cm³ and the 0.010-mol dm⁻³ aqueous silver nitrate titre was 23.45 cm³, what was the chloride ion concentration?

$$Ag^+(aq) + Cl^-(aq) \rightarrow AgCl(s)$$

_____
_____
_____
_____ (3)

4 Ammonium sulfate, (NH$_4$)$_2$SO$_4$, is used as a fertiliser. A 24.0-g sample of the fertiliser was added to excess NaOH(aq) and the resulting ammonia gas absorbed in 1-dm³ sulfuric acid, 0.10-mol dm⁻³. 10.0-cm³ of the resultant solution was titrated with 0.010-mol dm⁻³ NaOH. 29.10-cm³ was required. Calculate the purity of the ammonium sulfate.

$$(NH_4)_2SO_4(aq) + 2NaOH(aq) \rightarrow 2NH_3(aq) + Na_2SO_4(aq) + H_2O(l)$$

$$2NH_3(aq) + H_2SO_4(aq) \rightarrow (NH_4)_2SO_4(aq)$$

$$2NaOH(aq) + H_2SO_4(aq) \rightarrow Na_2SO_4(aq)$$

_____
_____
_____
_____ (3)

5 When 25.00-cm³ of a copper ion solution was mixed with excess potassium iodide solution, the iodine produced required 40.10-cm³ of 0.0200 mol dm⁻³ sodium thiosulfate solution. The reactions were these:

$$2Cu^{2+}(aq) + 4I^-(aq) \rightarrow 2CuI(s) + I_2(\text{in aqueous KI})$$

$$2S_2O_3^{2-}(aq) + I_2(aq) \rightarrow S_4O_6^{2-}(aq) + 2I^-(aq)$$

Calculate the concentration of the copper ion solution.

_____
_____
_____
_____ (3)

6 Benzene (M_r = 78) may be nitrated with concentrated nitric acid at 40 °C to make nitrobenzene (M_r = 123):

$$C_6H_6(l) + HNO_3(l) \rightarrow C_6H_5NO_2 + H_2O(l)$$

a A student used 10.0 grams benzene and obtained 9.0 grams nitrobenzene. What percentage yield was produced?

_____
_____
_____
_____ (1)

b Explain why the same student using the same amount could appear to make a yield of 120% later the same day (assuming the mathematics was correct).

_____
_____
_____
_____ (2)

7 During an exam practical or asessment, you were supplied with the solutions on the right:

1. Place P in a burette. Using a pipette, transfer 25.00-cm³ of solution Q into a conical flask. Using a measuring cylinder add 25-cm³ sulfuric acid. Warm by holding the flask over a Bunsen flame until the solution steams. Titrate the contents with solution P.

2. Using a pipette, transfer 25.00-cm³ of solution R into a conical flask. Using a measuring cylinder add 25-cm³ sulfuric acid and titrate the contents with solution P.

In Part 1, the average titre was 20.41 cm³. In Part 2, the titre was 12.55-cm³.

a Given that the following reaction takes place:

$$2MnO_4^-(aq) + 5C_2O_4^{2-}(aq) + 16H^+(aq) \rightarrow 2Mn^{2+}(aq) + 10CO_2(g) + 8H_2O(l)$$

calculate the concentration, in mol dm⁻³, of ethanedioic (oxalic) acid in solution Q.

_____
_____

b Using the titre in Part 2, calculate how many moles of potassium manganate(VII) (permanganate) was in the titre.

_____
_____

c How many moles of vanadium(III) chloride reacted with the permanganate ions in the titre?

_____
_____
_____

d What is the ratio between the moles of potassium manganate(VII) (permanganate) and vanadium(III) chloride?

_____
_____

e Write an equation for the reaction of vanadium(III) ions with the manganate(VII) (permanganate) ions, using the results of your calculations.

_____
_____

(Total 25 marks)

ANSWERS

UNIT 1

RECALL TEST

1. Electronic configuration.-(1)
2. Electronegativity increases across the row as proton number increases.-(2)
3. As atomic number increases within a group, electronegativity decreases as shells of electrons increase (so there is more shielding and a greater distance between the nucleus and the outer electrons). (3)
4. AgI is covalent because Ag and I are similar in electronegativity. (1)
5. Bonds are polar when the atoms in an covalent bond have different electronegativities. (1)
6. Van der Waals forces, permanent dipole or dipole–dipole, hydrogen bonding. (1)
7. The forces between the chlorine molecules are weak Van der Waals forces. (The covalent bonds are strong inside the molecule). (1)
8. As iodine atoms are large there are strong Van der Waals forces between the I_2 molecules. (1)
9. The Cl atom is more electronegative than the C atom. (1)
10. H bonds need an electron-deficient H (slightly positive), because the H is joined to a very electronegative atom, and N, O, or F atoms, which are small, very electronegative atoms with lone pairs that hold on to the H atoms. (4)
11. Ethanol has H bonds, but ethanal only has a permanent dipole. (2)
12. NaCl(l) has free ions. In NaCl(s) the ions are not free to move. (2)
13. The size of the atoms increase so the Van der Waals forces increase so the b.p. increases. (1)
14. Water H-bonds; the rest of the hydrides only have Van der Waals forces. (1)
15. The chlorine molecules lose kinetic energy, when the gas (separate randomly moving molecules) is cooled. Then the Van der Waals forces are strong enough to hold the chlorine molecules together in a liquid (touching randomly moving molecules). (4)
16. When solid NaCl (regular structure, ions touching) is heated the kinetic energy of the ions increases, until the electrostatic forces between ions is overcome, forming a liquid (ions randomly moving past each other, but touching). (4)

(Total 30 marks)

CONCEPT TEST

1. a Electronegativity is a measure of how attractive atoms are for a pair of electrons (in a covalent bond). (2)
   b Carbon dioxide has covalent bonding, a sharing of electrons. (2)
   c The -OH groups in glucose hydrogen-bond. Glucose dissolves in water because water can hydrogen-bond to the glucose. (2)
   d i Be^{2+} is much smaller than the Ca^{2+}.-(1)
      ii Aluminium chloride is covalent because the Al^{3+} ion is very small and highly charged so polarises the chloride ion. (2)
      iii The chloride ion is larger than the fluoride ion because the chloride ion has one more electron shell. (2)
      iv The F^- ion is very small and has a single charge so cannot be polarised by the aluminium ion. (1)
2. a Graphite is made of layers of covalently bonded carbon atoms. The layers are held together by weak Van der Waals forces, so the layers can slip over each other. (2)
   b All the atoms in diamond are strongly bonded together in one giant covalent lattice. (2)
   c C_{60} is a solid at room temperature, because the molecules are large so the Van der Waals forces between the molecules will be strong. As the particles would slide over each other it must be slippery. (2)
   d KC_{60}^+ ions with Cl^- ions must be ionic so have a high boiling point, because they are held together by strong electrostatic forces. (2)

(Total 20 marks)

UNIT 2

RECALL TEST

1. Neutron (no charge), proton (+). (If electrons are mentioned then no marks.) (2)
2. Vaporisation; ionisation; focus and acceleration; the magnetic field deflects the ions, separating them by mass and charge; the detector counts the ions of a particular mass/charge ratio; the vacuum pump removes the air to ensure ions are not deflected by air molecules. (6)
3. a Atoms of the same element (or atomic number) with different mass numbers (or numbers of neutrons). (2)
   b The weighted average mass of atoms of an element (in a sample of the element) divided by 1/12th of the mass of an atom of the carbon-12 nuclide. (2)
   c The number of protons and neutrons in the nucleus. (2)
4. The peak on the right. The one with the highest mass/charge ratio. (1)
5. Water: V-shaped, ammonia: pyramidal, methane: tetrahedral, beryllium chloride: linear, boron trifluoride: trigonal planar, sulfur hexafluoride: octahedral, phosphorus pentachloride: trigonal bipyramidal. (7)
6. Tetrahedral. 109.5°. (2)
7. Pyramidal. (1)
8. Methane = 109.5°, ammonia = 107°, water = 105°, carbon dioxide = 180°. (4)

9 The electron pairs (bond and lone pairs) repel until they are as far apart as possible. (1)
(Total 30 marks)

CONCEPT TEST

1. **a** i Using a magnetic field. (1)
   ii Without a vacuum the ions would be deflected by gas molecules and not get to the detector. (1)
   iii 107.972. (2)
   b i M_r = 60. (1)
   ii 15 = CH_3^+; 28 = CO^+; 45 = CO_2H^+; 60 = $C_2H_4O_2^+$ (lose 1-mark if no + charge on each ion). (4)
   iii CH_3COOH, $HOCH_2CHO$. (2)
   iv CH_3COOH because the mass/charge ratio of 45 is not possible for $HOCH_2CHO$. (2)
2. **a** 30 neutrons and 28 protons. (Maximum two marks if electrons are mentioned.) (3)
   b RAM = 58.768. (2)
3. **a** $AlCl_3$ has 3 bonds and no lone pairs, while NH_3 has 3-bonds and one lone pair. The electron pairs repel until they are as far apart as possible. (3)
   b i Draw pyramidal for PH_3 (it has 3 bonds and 1 lone pair).
   ii Draw a V-shaped molecule for SO_2 (it has 2 sets of double bonds and 1 lone pair).
   iii Draw pyramidal for ClO_3^- (Cl has 2 double bonds to Os, 1 single bond to an O^- ion, and has one lone pair).
   iv Draw a shape based on trigonal bipyramidal for BrF_3^- (the Br is surrounded by 3 bonds to F, and two lone pairs). You could arrange the lone pairs in any direction as long as the shape looks like it is based on trigonal bipyramidal. (4)
   c Octahedral. (1)
   d Bond angles: CH_4 =109.5°; NH_3 = 107°; H_2O = 105°. The-electron pairs repel until they are as far apart as possible. The lone pairs repel more than the bonding pairs, so with more lone pairs the bonding pairs are pushed together. (4)
(Total 30 marks)

UNIT 3

RECALL TEST

1. Magnesium atoms have one more proton than sodium atoms, and there are two electrons per atom involved in the metallic bonding compared with one electron per atom for sodium atoms. (2)
2. Mg colourless, Ca orange-red, Sr red, Ba pale or apple green, Li red, Na yellow, K lilac or pale purple. (7)
3. The potassium flame is visible through (cobalt) blue glass. (1)
4. Barium is less electronegative than magnesium. (1)
5. **a** $Mg(s) + 2HCl(aq) \rightarrow MgCl_2(aq) + H_2(g)$.
   b $MgO(s) + 2HCl(aq) \rightarrow MgCl_2(aq) + H_2O(l)$.
   c $MgCO_3(s) + 2HCl(aq) \rightarrow MgCl_2(aq) + CO_2(g) + H_2O(l)$.
   d $Mg(s) + Cl_2(g) \rightarrow MgCl_2(aq)$.
   e $2Mg(s) + O_2(g) \rightarrow 2MgO(s)$. (5)
6. The small double-charged Mg^{2+} ion polarises the peroxide anion, making it unstable. (2)
7. With increasing atomic number, the group 2 sulfates become less soluble. (1)
8. Because the group 2 ions become larger, so the hydroxide ion is less attractive, so the lattice energy decreases, making the hydroxides more soluble. (3)
9. Add aqueous barium chloride (or nitrate) and a heavy white precipitate will form which will not dissolve in dilute hydrochloric acid (or nitric acid), if sulfate ions are present. (2)
10. $Ca(OH)_2(aq) + CO_2(g) \rightarrow CaCO_3(s) + H_2O(l)$. (1)
11. The lithium ion is very small so polarises the nitrate anion, making it unstable. (2)
12. $BaCO_3(s) \rightarrow BaO(s) + CO_2(g)$. (1)
13. $2Na_2O_2(s) \rightarrow 2Na_2O(s) + O_2(g)$. (1)
14. $BaO_2(s) + H_2O(l) \rightarrow Ba(OH)_2(aq) + H_2O_2(aq)$. (1)
(Total 30 marks)

CONCEPT TEST

1. **a** When the sodium atoms are heated the electrons are promoted to higher energy levels. When the electrons fall back to lower energy levels visible light is emitted. (3)
   b Caesium reacts more vigorously than sodium, making the direct combination of Cl_2 with Cs dangerous. The Cs would react with the air. (2)
   c $NaCl(s) + (aq) \rightarrow Na^+(aq) + Cl^-(aq)$. (1)
   d There is a greater difference between the electronegativities of Na and I than there is between Li and I. (3)
2. **a** The group 2 sulfates become less soluble as the atomic number increases, because the cation radii increase so the cations become less attractive to the water molecules, so the hydration energy decreases. (The large sulfate anion means that the lattice enthalpy of the sulfates hardly differ.) (4)
   b Sulfate anion, SO_4^{2-}. (1)
   c As the atomic number increases the cation radii increase so the lattice energy decreases, as the larger cations keep apart the small fluoride anions. (3)
   d The magnesium ions are larger than the barium ions, so the Mg^{2+} ions polarise the carbonate anions more than the Ba^{2+} ions. (2)
   e The Al^{3+} ions have a greater charge and are smaller than the Mg^{2+} so polarise the CO_3^{2-} anion more, so carbonates will decompose at lower temperatures. (1)
(Total 20 marks)

UNIT 4

RECALL TEST

1. F_2 yellow, Cl_2 green, Br_2 red, I_2 black. (2)
2. Iodine molecules are large so the Van der Waals forces are strong. Chlorine molecules are small so the Van der Waals forces are weak. (2)
3. The atoms are small with few electron shells so the attraction between the nuclei and the shared electrons is strong. (2)
4. Chlorine is very reactive because of the energy released when forming ionic bonds, as the Cl electron affinity is very high, and the formation of Cl covalent bonds with other atoms is very exothermic. Iodine atoms are larger than chlorine atoms so the iodine electron affinity and covalent bonds release less energy. Iodine forms iodide ions which are large, releasing smaller lattice energies. (2)
5. a $2Fe(s) + 3Br_2(l) \rightarrow 2FeBr_3(s)$.
   b $H_2(g) + Br_2(g) \rightarrow 2HBr(g)$. (2)
6. Sodium *hydrogen* sulfate, hydrogen chloride. (2)
7. Name *or* formula. I_2, $KHSO_4$, KI, S, H_2S, SO_2. (3)
8. MgO +2, SO_2 +4, H_2SO_3 +4, SO_3 +6, H_2SO_4 +6, $MgSO_4$ +6, H_2S −2, NH_3 −3, NH_4^+ −3, $Na(s)$ 0, $Cl_2(g)$ 0. (11)
9. a $2Br^-(aq) + Cl_2(aq) \rightarrow Br_2(l) + 2Cl^-(aq)$.
   b No reaction.
   c $Ag^+(aq) + Cl^-(aq) \rightarrow AgCl(s)$.
   d $PCl_5(s) + H_2O(l) \rightarrow POCl_3(aq) + 2HCl(g)$ or $PCl_5(s) + 4H_2O(l) \rightarrow H_3PO_4(aq) + 5HCl$. (4)

(Total 30 marks)

CONCEPT TEST

1. a i Aqueous brine. (1)
      ii $2Cl^-(aq) \rightarrow Cl_2(g) + 2e^-$. (2)
   b Disproportionation is the simultaneous oxidation and reduction of the same element. (2)
   c $3Cl_2(g) + 6OH^-(aq) \rightarrow ClO_3^-(aq) + 5Cl^-(aq) + 3H_2O(l)$. Cl_2 0; ClO_3^- +5; Cl^- −1. (4)
   d $ClO_3^- + 2OH^- \rightarrow ClO_4^- + H_2O(l) + 2e^-$. (2)
   e Chloride ions form a white precipitate with aqueous silver nitrate, which dissolves in dilute ammonia. (2)
2. a $2S_2O_3^{2-}(aq) + I_2(aq) \rightarrow S_4O_6^{2-}(aq) + 2I^-(aq)$. (2)
   b Starch makes a dark colour with iodine which otherwise would be very pale near the end point. (2)
   c Add aqueous silver nitrate; a pale yellow precipitate would appear with I^-. (1)
   d The green gas becomes colourless, and a orange/brown solution or a black solid forms. (2)
3. a i $H_2SO_4(aq) + KCl(aq) \rightarrow KHSO_4(aq) + HCl(g)$. (1)
      ii An acid-base reaction. A proton (H^+ ion) is added to the chloride ion. Misty fumes are observed. (2)
   b i The solution turns red-brown because bromine is formed. (1)
      ii This is redox reaction. Sulfuric acid is not a strong enough oxidising agent to oxidise chloride ions, but it is strong enough to oxidise bromide ions. (2)
      iii S in sulfuric acid is +6. In sulfur dioxide it is +4. (1)
   c i A black solid forms (iodine). Also a 'bad-egg smell' is produced (hydrogen sulfide). (1)
      ii Sulfur and hydrogen sulfide (1)

(Total 30 marks)

UNIT 5

RECALL TEST

1. An 'atomic orbital' is the volume around a nucleus which is occupied by an electron 95% of the time. (2).
2. a to d See text. (4)
3. $Kr = 1s^2\ 2s^2\ 2p^6\ 3s^2\ 3p^6\ 3d^{10}\ 4s^2\ 4p^6$. (2)
4. An s-block element is one where the last electron added is to an s-orbital. (1)
5. The electronic configuration. (1)
6. The proton number increases across the row of the periodic table. (1).
7. The proton number increases across the row. (1)
8. a Increases.
   b Decreases.
   c Decreases. (3)
9. 1st ionisation energy: $X(g) \rightarrow X^+(g) + e^-$
   The energy required to remove one mole of electrons from one mole of gaseous atoms to form one mole of gaseous ions with a single positive charge. (1)
10. In the 2nd ionisation energy the electron is lost is from a 2+ ion, whereas the electron lost in the 1st ionisation energy is from a 1+ ion. (2)
11. The energy change when one mole of gaseous 3+ ions loses one mole of electrons to form one mole of gaseous 4+ ions. (2)
12. a He. (1)
    b He. (1)
    c Al or B. (1)
    d F. (1)
13. The metals Na, Mg, and Al have high melting points due to the metallic bonding (2). Si has giant covalent structure so the highest melting point (2), and the others form small molecules which are held together by weak Van der Waals forces. (2)

(Total 30 marks)

CONCEPT TEST

1. a $Cl(g) \rightarrow Cl^+(g) + e^-$. (2)
   b The electron lost from a He atom is from an unshielded shell (1) and the electron is held by two protons in the nucleus. (1)
   c 1st IE of S: $S(g) \rightarrow S^+(g) + e^-$
     2nd IE of S: $S^+(g) \rightarrow S^{2+}(g) + e^-$. (2)

d The group 2 elements become more reactive down the group. When the metals react they lose electrons. The 1st and 2nd IE decrease down group 2, so the electrons are lost from the atoms more easily down the group. (3)

2 a C has higher IE; C is smaller/has fewer shells than Si.
  b Ar has a higher IE; the electron is lost from a shell closer to the nucleus.
  c Be has higher IE; the electron lost from Be is from an s-orbital, closer to the nucleus than the electron lost from the p orbital in B.
  d Mg has higher IE; Mg has one more proton than Na.
  e Na has higher IE; the electron is lost from a shell closer to the nucleus. (10)

3 a The graph should show the IE generally increases as electrons are lost, with jumps after the 1st, 9th, and 17th electrons are lost. (4)
  b The electron is lost from a cation with a greater charge. (1)
  c The graph reflects the electronic structure of a potassium atom: 2, 8, 8, 1. The first e⁻ is lost from the outer shell, then the next 8 from one shell in, which is closer to the nucleus so the IE will be larger. The next 8-e⁻ is from the next shell in. The last 2 e⁻ are in the innermost shell so require the largest amount of energy. (4)
  d K has only 19 electrons. (2)

(Total 30 marks)

UNIT 6

RECALL TEST

1 Organic molecules that have the same functional group and react similarly form a homologous series. (1)
2 Methane CH_4, ethane C_2H_6, propane C_3H_8, butane C_4H_{10}, pentane C_5H_{12}, hexane C_6H_{14}, heptane C_7H_{16}, octane C_8H_{18}, nonane C_9H_{20}, decane $C_{10}H_{22}$. (3)
3 The strong covalent bonds in alkanes result in them being unreactive. The (C-C) and (C-H) bonds are strong (so have a high bond enthalpy – see unit 11), because the atoms are very small and unshielded, so the nuclei are held very strongly by the shared electrons. (1)
4 In the double bond the electrons in the pi bond are further from the nuclei than in a sigma bond, resulting in a weaker bond. Also the pi bond sticks out so is easily attacked by an electrophile. (1)
5 The nucleus of the large halogen atom is far from the shared electrons in the carbon–halogen bond. The electrons are also shielded by inner shells. (1)
6 a UV light or high temperature.
  b Room temperature.
  c Heat under reflux. (3)
7 a Structural isomers have the same molecular formula but different structural formulae. (1).
  b Geometric isomers have the same molecular structure but differ in arrangement in space by having two different groups on the same side (*cis*) or on opposite sides (*trans*). (1)
8 5 (including cyclobutane and methylcyclopropane). (1)
9 9 structural and geometric isomers. (1)
10 a 4.
   b 3.
   c 2.
   d 1. (4)
11 From C-F to C-I the bonds become weaker because the atoms are becoming larger so there is an increased distance between the nuclei and the bonding electrons, and there is increased shielding. (2)
12 (Induced Van der Waals = VdW, permanent dipole =-PD, hydrogen bonds = H) alkanes VdW, alkenes VdW, halogenoalkanes PD, alcohols H, aldehydes PD, ketones PD, amines H, nitriles PD, carboxylic acids H, carboxylic salts ionic, esters PD, amides H, carbonyl chlorides PD. (7)
13

graphical formula:

H H H
| | |
H—C—C—C—O—H
| | |
H H H

linear abbreviated formula: $CH_3CHOHCH_3$

skeletal formula: ⌳OH

(3)
(Total 30 marks)

CONCEPT TEST

1 a $C_6H_{12} + Br_2 \rightarrow C_6H_{12}Br_2$. (1)
  b The bonds in hexane are too strong to be broken by bromine, because the atoms in hexane are very small. UV light is required to initiate the reaction. (2)
2 a Room temperature.
  b Heat under reflux.
  c Ultraviolet light.
3 a The bonds become weaker because the atoms are becoming larger so there is an increase distance between the nuclei and the bonding electrons, and there is increased shielding. (3)

b The C=C bond is stronger than the C-C bond because there are more electrons involved in the C=C double bond. The C=C is less than double the strength of the single C-C covalent bond because both contain a single sigma bond, but the second bond in the C=C bonding is a pi bond in which the electrons are further from the nucleus than in a sigma bond. (3)

c The Si-Si bond is weaker because the Si atoms are larger so the electrons in the bond are further from the Si nucleus. (3)

4 a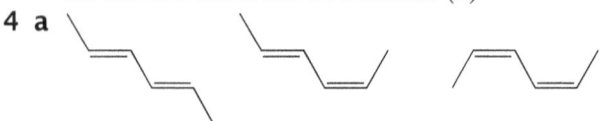

b The C=C make this molecule very reactive due to the high electron density in the C=C. This makes the molecule very attractive to electrophiles. (5)

(Total 20 marks)

UNIT 7

RECALL TEST

1 a Alkali.
  b Acid.
  c Oxidising agent.
  d Reducing agent.
  e Reagent that causes hydrolysis.
  f Dehydrating agent. (6)

2 a Condensation/addition elimination. (1)
  b Alkaline hydrolysis.-(1)

3 a A molecule or ion with an electron-rich site which can donate a pair of electrons. (2)
  b A molecule or ion with an electron-deficient site which can accept a pair of electrons. (2)
  c When a group of atoms is added to a molecule, and no atoms are lost. (2)
  d When an atom or one group of atoms is replaced by another group.-(2)

4 a Nucleophilic substitution. (2)
  b Electrophilic addition. (2)

(Total 20 marks)

CONCEPT TEST

1 a Oxidising agent: $KMnO_4$(aq) with H_2SO_4(aq)/$K_2Cr_2O_7$(aq) with H_2SO_4(aq).
  Reducing agent: $LiAlH_4$(dry ether)/$NaBH_4$(aq)/H_2(g) with Ni(s) or Pt(s).
  Dehydrating agent: concentrated H_2SO_4/solid P_4O_{10}. (3)

  b When CH_3CH_2OH is converted to CH_2CH_2 then two H-atoms and one O atom are lost: H_2O is lost, so the change is called dehydration. (1)

  c Any sodium-containing base: NaOH(aq), or $NaHCO_3$(s), or Na(s). (1)

  d Any strong acid: H_2SO_4(aq), or HCl(aq). (1)

2 a Condensation/addition elimination/esterification. (1).

b Water, as the reaction is hydrolysis. (1)

3 a A molecule or ion with an electron-rich site which can donate a pair of electrons replaces a group of atoms.(2)
  b When bonds break and the pair of electrons go with one atom. (2)
  c See Fig 8.5. You should draw the nucleophile as a CN^--ion rather than a OH^- ion. The lone pair is on the C in CN^-. (3)

4 a See Fig 8.6. Instead of the H-Br there is Br-Br, so a Br joins to the ethene first (instead of the H). (3)
  b Bromine reacts by electrophilic addition, whereas the C=O reacts by nucleophilic addition. The Br_2 attacks the C=C to make Br-C-C$^+$. A C=O group cannot react in the same way, as the O atom cannot have a positive charge. Also, if a C$^\oplus$ formed that would mean a weak O-Br bond would have to form also. (2)

(Total 20 marks)

UNIT 8

RECALL TEST

1 Chlorine and ultraviolet light. (2)
2 More chlorine with ultraviolet light. (2)
3 A molecule or atom with an unpaired electron. (2).
4 Initiation: $Br_2 \rightarrow 2Br^{\bullet}$(g) with UV light.
  Propagation: $CH_3CH_3 + Br^{\bullet} \rightarrow {}^{\bullet}CH_2CH_3 + HBr$ (remember this one), then ${}^{\bullet}CH_2CH_3 + Br_2 \rightarrow CH_3CH_2Br + Br^{\bullet}$.
  Termination: ${}^{\bullet}CH_2CH_3 + Br^{\bullet} \rightarrow CH_3CH_2Br$, or ${}^{\bullet}CH_2CH_3 + {}^{\bullet}CH_2CH_3 \rightarrow CH_3CH_2CH_2CH_3$ or on the reaction vessel walls $Br^{\bullet} + Br^{\bullet} \rightarrow Br_2$. (6)
5 By fractional distillation. (2)
6 To make useful alkenes and to turn cheap long-chain alkanes into short-chain alkanes which are worth more. (2)
7 Electrophilic addition. (1)
8 Room temperature. (1)
9 a $CH_2CH_2 + HBr \rightarrow CH_3CH_2Br$.
  b $CH_2CH_2 + Br_2(CCl_4) \rightarrow CH_2BrCH_2Br$.
  c $CH_2CH_2 + Br_2$(aq) $\rightarrow CH_2BrCH_2OH$ (mixed with CH_2BrCH_2Br).
  d $CH_2CH_2 + H_2O \rightarrow CH_3CH_2OH$. (4)
10 Using phosphoric acid, H_3PO_4, and steam (1)
11 Hydrogen gas with nickel catalyst and 200 °C temperature. (3)
12 Alkenes decolorise bromine water. (1)
13 -[-$CH_2C(CH_3)H$-]-. (1)
14 Microbes cannot decompose them because they are chemically inert and waterproof. (2)

(Total 30 marks)

CONCEPT TEST

1 a Chlorine and ultraviolet light. (2)
  b Heat under reflux with conc. sulfuric or phosphoric acids, then dilute with water, or pass ethene and steam over an acid catalyst. (2)

c Add bromine to make dibromoethane. Next add aqueous NaOH, wa=heat under reflux, to make ethane-1,2-diol. (2)

d Add hydrogen gas, Ni (or Pt) catalyst, 200 °C and high pressure. (2)

2 a Reagent: bromine water. (1)
Conditions: room temperature. (1)
Observation with palm oil: no change. (1)
Observation with sunflower oil: bromine water is decolorised. (1)

b Advantage: decomposes when discarded. Disadvantage: object could break down before it is finished with. (2)

3 a A molecule or atom with an unpaired electron. (2)

b The ultraviolet light in sunlight splits bonds to make free radicals. (1)

c Initiation: ultraviolet light + $Cl_2 \rightarrow 2Cl^\bullet$.
Propagation: $-CH_2- + Cl^\bullet \rightarrow -^\bullet CH- + HCl$, then $-^\bullet CH- + Cl_2 \rightarrow -CHCl- + Cl^\bullet$, and Termination: $^\bullet CH- + Cl^\bullet \rightarrow -CHCl-$. (3)

(Total 20 marks)

UNIT 9

RECALL TEST

1 Nucleophilic substitution. (2)
2 **a** $CH_3CH_2Cl(l) + NaOH(aq) \rightarrow CH_3CH_2OH(aq) + NaCl(aq)$.
   b $CH_3CH_2I(l) + 2NH_3(ethanol) \rightarrow CH_3CH_2NH_2 + NH_4I$.
   c $CH_3CH_2Br(l) + KCN(ethanol) \rightarrow CH_3CH_2CN + KBr$.
   d $CH_3CH_2Br + KOH(ethanol) \rightarrow CH_2CH_2 + H_2O + KBr$. (8)
3 CH_3CH_2OH, ethanol. (2)
4 Ethanolic KCN, boil under reflux. (2)
5 Aqueous hydroxide ions with bromoethane produces ethanol. Hydroxide ions in ethanol produces ethene. (4)
6 Silver bromide, AgBr. It must be a bromoalkane. (2)
7 Haloalkanes do not have strong enough dipoles to attract water molecules. (2)
8 Chloroethane is a much larger molecule)than propane. Propane only has Van der Waals forces between the molecules. Chloroethane contains a polar bond so there is also dipole-dipole attraction between the chloroethane molecules. (2)
9 **a** fluoromethane,
   b bromoethane
   c 1,2-dibromoethane,
   d 1-bromo 2,3-dichlorobutane (8)
10 **a** O_3,
   b chlorfluorocarbon molecules.
   c $O_2 + O \rightleftharpoons O_3$
   d ultra- violet light
   e $O_3 + Cl^\bullet \rightleftharpoons O_2 + OCl^\bullet$, then $OCl^\bullet + O_3 \rightleftharpoons 2O_2 + Cl^\bullet$.
   f hydrofluorocarbons, carbon dioxide (8)

(Total 40 marks)

CONCEPT TEST

1 **a** $CH_3CH(CH_3)CH_2Br$.
  b $CH_3CH_2CH(Br)CH_3$. (2)
2 **a** Aqueous NaOH produces propan-2-ol. (2)
     NaOH in ethanol produces propene. (2)
  b **i** NH_3(ethanol), heat under reflux. (2)
     ii KCN(ethanol), heat under reflux.(2)
  c Add aqueous silver nitrate. A cream precipitate forms if-a C-Br group is present. (2)
3 **a** $CH_3CH_2CH_2CH_2Br$ or $CH_3CH(CH_3)CH_2Br$.
  b $CH_3CH_2CHBrCH_3$, also written $CH_3CHBrCH_2CH_3$.
  c $(CH_3)_3CBr$. (3)
4 **a** A: aqueous NaOH or KOH; heat under reflux.
     B: ethanolic NaOH or KOH; heat.
     C: ethanolic KCN; heat. (6)
  b $CH_3CH(NH_2)CH_2NH_2$. (2)
  c To help you: 2,2-dimethyl 1-chloropropane is $(CH_3)_3CCH_2Cl$.
     i It does not react.
     ii $(CH_3)_3CCH_2OH$. (2)

(Total 25 marks)

UNIT 10

RECALL TEST

1 Oxidation, substitution, dehydration, esterification (addition-elimination), reduction (when Na is used). (4)
2 Short-chain alcohols hydrogen-bond with water, while long-chain alcohols are more like alkanes so held mostly by Van der Waals forces. (3)
3 Ethanol has hydrogen bonds, ethanal has only a permanent dipole. (2)
4 **a** Ethanol or hydroxyethane.
  b Propan-2-ol.
  c 2-methylpropan-2-ol.
  d 2,2-dimethylpropan-1-ol. (8)
5 **a** $CH_3CH_2OH + Na \rightarrow CH_3CH_2O^-Na^+ + \frac{1}{2}H_2$.
  b $CH_3CH_2OH + 2[O] \rightarrow CH_3COOH + H_2O$.
  c $CH_3CH_2OH \rightarrow CH_2CH_2 + H_2O$.
  d $CH_3COOH + CH_3CH_2OH \rightarrow CH_3COOCH_2CH_3 + H_2O$.
  e $CH_3CH_2OH + PCl_5 \rightarrow CH_3CH_2Cl + POCl_3 + HCl$.
  f $CH_3CH_2OH + HBr \rightarrow CH_3CH_2Br + H_2O$.
  g $CH_3CHOHCH_3 + [O] \rightarrow CH_3COCH_3 + H_2O$.
(8)

CONCEPT TEST
1   a $CH_3CH(CH_3)CH_2OH$.
    b $CH_3CH_2CH(OH)CH_3$. (2)
2   a Heating under reflux produces propanoic acid. (2)
       Distilling the mixture produces propanal (2)
    b i sodium metal. (2)
       ii KCN(ethanol), heat under reflux. (2)
    c Add phophorous pentachloride and a white fumes are made; OR add ethanoic acidd and drops of sulfuric acid and detect fruity smell; OR Use infra-red spectroscopy and state the required wave number rage (see the main text for the wave number range) (2)
3   a $CH_3CH_2CH_2CH_2OH$ or $CH_3CH(CH_3)CH_2OH$.
    b $CH_3CH_2CHOHCH_3$, also written $CH_3CHOHCH_2CH_3$.
    c $(CH_3)_3COH$. (3)
4   a A: ethanoic acid; conc. H_2SO_4; heat under reflux.
       B:-conc. HBr (or KBr and H_2SO_4); heat under reflux.
       C: concentrated H_2SO_4; or aluminium oxide; or pumice; heat. (6)
    b $CH_3COOCH(CH_3)_2$. (2)
    c To help you: 2-methylpropan-2-ol is $(CH_3)_3COH$.
       i   Methylpropene $CH_2CH(CH_3)CH_3$.
       ii It does not react. (2)

(Total 25 marks)

UNIT 11

RECALL TEST
1 Exothermic reactions have a negative enthalpy. (1)
2 a The enthalpy change when one mole of a substance is formed from its constituent elements in their normal states (under standard conditions). (3)
   b The enthalpy change when one mole of a substance is completely combusted in oxygen (under standard conditions). (3)
   c The enthalpy change when one mole of water is formed by reacting an acid and base (under standard conditions). (3)
3 The energy required to break one mole of a particular kind of bond in a particular compound. (3)
4 The average energy required to break one mole of a particular kind of bond taken from many compounds. (3)
5 The energy change accompanying a reaction is independent of the route taken. (2)
6 Quantity of heat (q) = mass (m) × specific heat capacity (c) × temperature change (ΔT). (2)

(Total 20 marks)

CONCEPT TEST
1 a +175 kJ mol^{-1}. (4)
   b +35 kJ mol^{-1}. (4)
   c ii –186 kJ mol^{-1}. (3)
2   9.7 kJ mol^{-1}. (2)
3 a –1169 kJ mol^{-1}. (3)
   b 391 kJ mol^{-1}. (4)

(Total 20 marks)

UNIT 12

RECALL TEST
1 Rate is the measure of how fast a reactant (or product) concentration changes over time. The units of rate are usually mol dm^{-3} s^{-1}. (2)
2 Temperature, concentration, pressure, catalyst, surface area, light. (6)
3 Collisions with each other, with the correct orientation, and with enough energy. (3)
4 When the concentration is increased there are more molecules in the same volume, so they collide and react more often, and the rate is increased. (2)
5 When the surface area is increased there is more contact between the molecules so the rate increases. (2)
6 The diagram is like Fig. 12.3, but with the reactants lower than the products. Also the profile includes a low-energy state between two transition states. (2)
7 Activation energy is the minimum energy required in a molecular collision for the molecules to react. (3)
8 See Fig. 12.5. Did you ensure the areas under the curves are similar and that the higher temperature peak is lower than the lower temperature? Activation energy must be to the right of the peaks. (3)
9 When the temperature increases the molecular kinetic energy increases, so there are more molecules colliding with an energy greater than the activation energy. This is shown by the increase in the shaded area on the graph. (2)
10 A catalyst increases the reaction rate without being used up. (1)
11 A catalyst increases the reaction rate by creating an alternative reaction route with a lower activation energy. (2)
12 Catalysts produce an alternative reaction route by having a variable oxidation state, or by acting as a surface that adsorbs molecules, bringing them closer together. (2)

(Total 30 marks)

CONCEPT TEST
1 a The minimum energy required in a molecular collision for the molecules to react. (3)
   b The diagram is like Fig. 12.3, but with the reactants lower than the products. Also the profile includes a low-energy state between two transition states. (3)
   c See Fig. 12.5. (3)

d When the temperature increases, the molecular kinetic energy increases, so there are more molecules colliding with an energy greater than the activation energy. More molecules react, so the reaction rate increases. (3)
   e Rate. (1)
2 **a** See Fig. 12.6. (2)
   b A catalyst increases the reaction rate by creating an alternative reaction route with a lower activation energy so that there are more molecular collisions with an energy greater than the activation energy, so more molecules react. (3)
   c Iron is a heterogenous catalyst because it is a different state (solid) from the reactants (gases). (2) (Total 20)

UNIT 13

RECALL TEST

1 'Dynamic equilibrium' is when in a reversible reaction the concentrations of reactants and products do not change, but the reactants are continually producing products and the products produce reactants. (1)
2 Le Chatelier's principle states that if the conditions of a system at equilibrium are changed then the equilibrium position will shift to resist the change. (1)
3 **a** Right.
   b Left.
   c Left.
   d Does not change (as the reaction ΔH is zero).
   e Does not change (as catalysts do not change the position of equilibrium). (5)
4 **a** Left.
   b Left.
   c Right.
   d Does not change the pressure of the product. (4)
5   forward backward yield
   a incr. temperature,   I   I   D
   b incr. pressure, I   I   No
   c add catalyst. I   I   No
   (I = increase, D = decrease, No = no change.) (6)
6 **a** $N_2(g) + 3H_2(g) \rightleftharpoons 2NH_3(g)$.
       Catalyst: iron. 450 °C and 200–1000 atm.
   b $SO_2(g) + \frac{1}{2}O_2(g) \rightarrow SO_3(g)$.
       Vanadium(V) oxide at 450 °C and 1–2 atm.
   c $4NH_3(g) + 5O_2(g) \rightleftharpoons 4NO(g) + 6H_2O(g)$.
       Pt/Rh catalyst at 850 °C and 1–2 atm. (3)
       (Total 20 marks)

CONCEPT TEST

1 **a** If the temperature increased the yield would decrease, because an increase in temperature would shift the reaction position of equilibrium to the left as the forward reaction is exothermic. (2)
   b High pressure would be very expensive, requiring thick-walled pipes and compressors. The yield is most economic at the stated temperature. (2)
   c The catalyst would increase the rate of production of ammonia by obtaining the yield sooner as the rate increases. (2)
   d The catalyst would not change the yield because catalysts do not influence yield, only rate. (2)
2 **a** 'Dynamic equilibrium' is when in a reversible reaction the concentrations of reactants and products do not change, but the reactants are continually producing products and the products produce reactants. (2)
   b Increasing the temperature would shift the reaction equilibrium to the left, as the forward reaction is exothermic. (2)
   c No, because there are more gas molecules on the product side of the reaction so an increase in pressure would shift the position of equilibrium to the side with fewer molecules (so resisting the increase in pressure slightly). (2)
   d The yield is economic at low pressure. Higher pressure would be expensive and would decrease the yield. (2)
   e $NO(g) + O_2(g) \rightarrow NO_2(g)$, then $2NO_2(g) + H_2O(l) + \frac{1}{2}O_2(g) \rightarrow 2HNO_3(aq)$ (Other equations are possible. In some plants excess NO_2 is distilled off). (2)
   f $2NH_3 + H_2SO_4 \rightarrow (NH_4)_2SO_4$ (2)
       (Total 20 marks)

UNIT 14

RECALL TEST

1 Kinetics, equilibrium, enthalpy, economic, and environmental factors. (5)
2 Temperature, concentration, pressure, catalysts, surface area (or light). (5)
3 Temperature, concentration, pressure, and economic factors. (4)
4 Low. (1)
5 Both increase costs unless near-atmospheric conditions. (1)
6 $SO_2(g) + \frac{1}{2}O_2(g) \rightleftharpoons SO_3(g)$. (1)
7 Both yield and rate would decrease. (1)
8 Increased yield. Decreased rate. (1)
9 Increased rate. No effect on yield. (1)
10 1–2 atm pressure, 450 °C, and V_2O_5 catalyst. (2)
11 SO_3 is dissolved in pure H_2SO_4 which is then diluted. (1)
12 $4NH_3(g) + 5O_2(g) \rightleftharpoons 4NO(g) + 6H_2O(g)$. (1)
13 $2NO(g) + O_2(g) \rightarrow 2NO_2(g)$. (1)
14 Fertilisers, explosives, and polyamides. (3)
15 Ammonium sulfate/ammonium nitrate. (1)
16 Lead compounds were added to limit pre-ignition. (1)

17 Lead harms the nervous system, and poisons catalytic converters. (2)
18 They convert pollutants (CO, NO_x, unburnt hydrocarbons) into CO_2, N_2, and H_2O. (3)
19 Rhodium, Rh. (1)
20 Catalytic poisons bind irreversibly with catalysts. (1)
21 Ethanol is made by hydrating ethene over a catalyst of phosphoric acid at 300 °C and 70 atm. (1)
22 Alcoholic drinks, fuels, solvents. (1)
23 Petrol additive and as an industrial feed stock. (1)

(Total 40 marks)

CONCEPT TEST

1 a Pre-ignition is when fuel combusts too early in a car engine. It causes damage to the car engine, loss of power, and increased fuel consumption. (2)
  b Otherwise methanol production would be too slow to be economic. (2)
  c Yield of methanol would increase/ equilibrium would shift to right. The rate would increase. (2)
  d Decrease the yield/equilibrium shifts to the left (exothermic reaction). Increase the rate. (2)
  e Improved combustion so less CO, NO_x, unburnt fuel. Less sulfur dioxide/acid rain. (2)
  f Lessen NO_x emissions, so less acid rain. Less poisonous CO and unburnt fuel. (3)
  g Heterogeneous, because the solid catalyst is in a different phase to the exhaust gases. (2)

(Total 15 marks)

UNIT 15

RECALL TEST

1 See Fig. 15.4. (2)
2 Add aqueous NaOH and the gas coming off should turn pink litmus blue. (3)
3 a Br^- ions.
  b Possibly CO_2.
  c A sulfite (a sulfate(IV) compound).
  d A sulfate (sulfate(VI) compound).
  e A nitrate. (5)

(Total 10 marks)

CONCEPT TEST

1 a The white solid must be dissolved in the minimum amount of hot solvent, filtered hot using vacuum filtration. The fitrate is cooled slowly, then filtered cold. The solid is the purified compound. (4)
  b Seal one end of a piece of capillary tube. Tap some of the purified solid into the tube. Attach the tube upright to a thermometer so that the bottom ends are together. Put into an oil bath. Warm it gently, while stirring, and note the temperature at which the crystals melt. Repeat until the readings are the same. (3)
  c Put some of the liquid in a flask. Clamp a thermometer so that the bulb is just above the liquid surface. Warm the liquid slowly until it is boiling. Repeat. (2)
2 a X: $BaNO_3$; Y: $BaSO_4$; Z: NO_2. (6)
  b P: K_2SO_3; Q: SO_2. (2)
  c Add aqueous silver nitrate, and nitric acid. A cream (off-white) precipitate would form that would dissolve in concentrated ammonia solution. (3)

(Total 20 marks)

UNIT 16

RECALL TEST

1 a mole = mass/RAM.
  b mass = mole × RAM.
  c RAM = mass/mole. (6)

2 a i 72 dm^3.
    ii 12 dm^3. (2)
  b 41.7 moles. (1)

3 density = mass in grams /volume in cm^3. (1)
(Total 10 marks)

CONCEPT TEST

1 a 84 / 24 000 = 0.0035 moles
  b 0.0035 moles
  c formula mass = mass / moles = 140
  d 140 = (2 x RAM) + 12 + (3 x 16)
    So RAM = 40. so the group 1 metals must be potassium. Note the slight experimental error. (5)
2 a Reaction 1: 23 %
    Reaction 2: 61 %
  b The second reaction has the better atom economy so is the better process. (5)

(Total 10 marks)

UNIT 17

RECALL TEST

1 a concentration = mole/volume (in dm^3).
  b moles = concentration × volume (in dm^3).
  c volume (in dm^3) = moles/concentration. (6)
2 a mole = mass/RAM.
  b mass = mole × RAM. (3)
  c Percentage yield = $\frac{\text{ACTUAL moles}}{\text{POSSIBLE moles}}$ × 100%. (2)

(Total 10 marks)

CONCEPT TEST
1. $0.256 \text{ mol dm}^{-3}$. (2)
2. $I_2(\text{in KI(aq)}) + 2S_2O_3^{2-}(\text{aq}) \rightarrow 2I^-(\text{aq}) + S_4O_6^{2-}(\text{aq})$.
   Concentration iodine = $0.00538 \text{ mol dm}^{-3}$. (3)
3. Concentration chloride ions = $0.0094 \text{ mol dm}^{-3}$. (3)
4. 94.0% pure. (3)
5. $0.0321 \text{ mol dm}^{-3}$. (3)
6. a Percentage yield = 57%. (1)
   b Dinitrobenzene was made, probably due to the temperature rising above the 60-°C required for nitration. (2)
7. a Conc. $H_2C_2O_4$(aq) acid = $0.0408 \text{ mol dm}^{-3}$.
   b 0.000 251 moles.
   c 0.001 25 moles V^{3+} ions.
   d Ratio = 0.201 or 1/4.97.
   e Ratio rounds to 1:5 so $MnO_4^-(\text{aq}) + 5V^{3+}(\text{aq}) + 8H^+(\text{aq}) \rightarrow Mn^{2+}(\text{aq}) + 5V^{4+}(\text{aq}) + 4H_2O(l)$. (8)

(Total 25 marks)

INDEX

absorbances in infrared spectra 58
acid
  reactions 15, 38
acid–base reactions 15, 38
activation energy 68
addition 38, 39, 46
  polymer 46
alcohol 33, 38, 44, 51, 56
  primary, secondary, tertiary 56
alkali metals 14
alkaline earth metals 14
alkanes 32, 33, 44, 45, 46
  general formula 32
alkenes 32, 33, 44, 45, 46, 51
  general formula 45
aluminium
  manufacture 28, 81
amide 33
amine 33, 38, 51
ammonia 21, 51, 74, 82, 87
  production, uses 74
amount of substance 94
analysis 87, 88
anion 2
anionic
  radius 28
anode 21
aromatic 39, 45
atomic
  number 2, 28
  orbital 25
  radius 16, 28
Avogadro constant 93

barium 14, 15, 16
base reactions 15
bauxite 81
beer 57
benzene 34, 45
beryllium 14, 15, 16
biofuels 44, 56
bitumen 44, 45
blast furnace 80, 81
bleach 4, 20, 21
boiling point 3, 4, 26, 32, 56
  determining 86
bond
  angles 8
  enthalpy 32, 62
  strain 46
bonding molecular orbital 26
bromide 21, 22
bromination 40
bromine 20, 21, 22, 32
  extraction of 15
bromopropane isomers 34
Buchner funnel 87

but-2-ene isomers 34
butane 4, 32, 34, 44

calcium 14, 15
  hydroxide 16, 17
carbon 7
  dioxide 8, 15, 16
carbonate 15, 16, 38, 87
  thermal decomposition 16
carbocation (carbonium ion) 45
carbonium ion (carbocation) 45
carbonyl 33
  compounds 78, 79
carboxylic acid 33, 57, 58
catalysts 45, 46, 52, 68, 70, 76, 80, 81, 82
  heterogeneous 70
  homogeneous 70
catalytic converters 44
cathode 21, 81
cation 2
cationic radius 16, 28
CFCs 52
chain reaction 40
chemical
  calculations 92, 98
  properties, determination 2, 20
chlorate(I) ion 21, 22
chlorate(V) ion 21
chloric(I) acid 21
chloride 14, 15, 22, 98
chlorine 3, 15, 20, 21, 22, 52, 81
  production 21
chloro 50, 51, 52
chloroethene 46
chlorofluorocarbons (CFCs) 52
cis (E-Z isomerism) 34
closed system 74
co-ordinate bond 3
colour 15, 20, 21, 45, 87
  of aqueous cations 87
coloured flames 15, 87
compounds
  intermediate 45, 64
  primary, secondary, tertiary 56
concentrated sulfuric acid 21, 80, 82, 99
concentration 98
condensation 38, 58
condenser 33, 50, 86
conductivity 28
Contact process 81, 82
covalent

bond 2, 3, 4, 26
character 14, 15
cracking 44, 45
curly arrow 40

dative covalent bond 3
decane 32
decimetre 98
dehydration 38, 56, 58
density 93
diaphragm cell 21
diol 48
dipole–dipole (permanent dipole) interactions 4, 50
displacement 19
disproportionation 21
dot and cross diagram 8
double bond 26, 32
dynamic equilibrium 74

E-Z isomerism (geometric) 34
economics of industrial processes 82
electrode 21, 81
electrolysis 80
  of alumimium oxide 81
  of brine (impure sodium chloride) 21
electron affinity 28, 64
electronegativity 2, 3, 14, 26
electronic configuration 26
electrophile 39
electrophilic addition 39, 45
elimination 38, 51
endothermic 62
energetics 62
energy barriers 68
enthalpy
  change 62, 63
  in experiments 63
  of combustion 62
  of formation 62, 44
  of neutralisation 62
environment 80
enzyme 81
epoxy resins 46
epoxyethane 46
equilibrium 74, 75, 76
  law 74
esterification 38, 56
esters 33, 38, 56
ethanal 4, 38
ethane 32, 70
ethanoic acid 38
ethanol 4, 38,
ethene 8, 38, 44
ethylamine 51
exothermic 62

experimental skills 86
explosives 82

fats and oils 82
fermentation 63
fertilisers 82
filtering under reduced
    pressure 87
filtrate 87
filtration 87
fish hook 40
flame colours 15
flooring 46
fluctuating dipole 4
fluorine 20, 22, 52
formula mass 4, 99
formulae
  graphical/display 33, 34
  linear 33, 34
  molecular 33, 34
  skeletal 34, 35
  structural 33, 34
fractional distillation 44, 46
fractions 44
free radical 40, 45, 52
  substitution 40
functional groups 33

gasohol 44
geometric isomerism (E-Z
    isomerism) 34
giant covalent lattice 3, 27
giant ionic lattice 2
giant metallic lattice 3
group 2
  1 (alkali metals) 14
  2 (alkaline earth metals) 14
  7 20

Haber process 74, 75
haematite 80
half equations 15
halide 20, 21, 87
halogen 20, 32, 50
halogenation 56
haloalkane 50, 51, 56
heat change calculations 63
heating under reflux 33, 50,
    57, 86
heptane 32
Hess's law 62
heterolytic fission 40
hexane 32
homologous series 32
homolytic 40
homolytic fission 40
hydration 38

hydrogen
  bonding 4, 51
  bromide 21, 38
  chloride 4, 21
  gas 15, 44, 45
  iodide 21
  peroxide 15, 22
  production 21
hydrogenation 45
hydrogencarbonate 38
hydrogensulfate 87
hydrolysis 38, 51
hydroxide 15, 16, 21, 51

incinerate 46
indicator 21
infrared
  absorption wave
    numbers 58
  spectroscopy 58
initiation 40
insecticide 52
intermolecular forces 4
iodide 21, 22, 70
iodine 21, 22, 57
ionic
  bond strength 3
  bonds 3
  character 3
  chlorides 15
  compounds 2, 3, 14, 15, 16
  lattice 2, 3
  oxides 15
ionisation energy 14, 26, 27,
    64
  of successive elements 21
iron 80, 81
  (II) ions as catalyst 70
  ore 80
isomerisation (reforming) 44,
    45
isomerism 26, 27, 86, 95
  E-Z isomerism 34
  geometric 34
  structural 34
isomers, branched 32, 44, 45
isotopes 9

ketone 57
kinetic stability 15, 16, 82
kinetics of reactions 68

Le Chatelier's principle 74
lead 44
  replacement fuel (LRF) 44
limestone 75
linear (shape) 8

liquid petroleum gas
    (LPG) 44
lithium 14
London forces 4
lone pair 4
LPG 44
LRF 44

macromolecular 2
magnesium 81, 14
manganese 80
margarine 45
mass
  number 9
  spectrometer 9
  spectrum 9
Maxwell–Boltzmann
    distribution 69
mean bond enthalpy 32, 62
mechanisms 39, 40
melting points 2, 3, 4, 14, 26
  determining 86
membrane cell 21
metals 2
  oxides 14
metallic bonds 3, 14
methane 32
methanol 44, 56, 76, 85
molar
  concentration 98
  gas volume 94
mole 92
molecular
  ion 10
  shape 8
monomer 46

natural gas 44
nickel
  catalyst 45, 70
nitrate 21, 22, 51, 82
nitric acid production 82
nitrile 32, 51
nitrogen dioxide 82
noble gases 26
nonane 32
non-metals 3
non-stick coating 46
nucleophile 39, 51
nucleophilic substitution 39
nucleus 2

[O] 57
orbital 26, 25
organic
  bonding 32
  mechanisms 39, 40

115

oxidation   14, 21, 22, 38, 56, 57
oxidation number/state   22
   variable   81
oxides   15, 80
oxidising agent   21, 38, 57
oxygen   15, 22, 38, 44, 81
ozone layer   52

paint,   82
parent ion   10
pentane   32
percentage yield   115
percentage atom economy   94
period   2
periodic table   2, 26
permanent dipole
   interactions   4
peroxides   15, 21
petrol   33, 44, 45, 56
phosphoric acid, catalyst   56, 81
phosphorus   57
   pentachloride   57
pi (π) bond   26, 32
pipes   46
plastic
   bags   46
   bottles   46
   recycling   46
platinum/rhodium
   catalyst   82
polar covalent bonds   4
polarisation (dipole–
   dipole)   4, 50
poly(chloroethene) (PVC)   46
poly(ethene)   46
polymers   46, 52
   addition   46
   biodegradable   46
poly(propene)   46
poly(tetrafluoroethene)
   (PTFE)   46
potassium   14
   dichromate   57
   permanganate   57, 99
practicals
   planning   88
   techniques   87
precipitation   16, 21, 51, 87
precision   87
pre-ignition   44, 45
product   74
propagation   40
propane   32
propanone   33, 57
propene   33, 46
PTFE   33
PVC   33

rate   68, 69, 70
   forward and backward   74
   in practicals   88
reactant   68
reaction
   mechanism   39, 40
   profile   68, 69
redox   22
reducing agent   22, 38, 80
reduction   21, 22, 80, 81
reforming   44, 45
relative abundance   10
relative atomic mass   9
relative isotopic mass   9
relative molecular mass   9
repeating unit   45
residue   87
reversible reactions   74

s block   14
safety   87
sea water   22
shape of molecules   8
   linear   8
   octahedral   8
   pyramidal   8
   square planar   8
   tetrahedral   8
   trigonal bipyramidal   8
   trigonal planar   8
   V   8
shell   2, 20
shielding   2
sigma (σ) bond   20
silicon   27
   dioxide   22
silver
   catalyst   46, 70
   nitrate   21, 22, 51
simple covalent
   bond   3
simple oxides   15
sodium   14, 22, 81
   hydroxide   15, 21, 51
   hydroxide production   21
   thiosulfate   21
solubility, general pattern   99
   of group 2 hydroxides   16
   of group 2 sulfates   16
specific heat capacity   63
spectroscopy   58
standard conditions   62
standard solution   98
strontium   14, 15
subshells   20
substitution   38, 39, 50, 57
sulfite   16, 87
sulfur   22
   dioxide   8, 80, 82

sulfuric acid   21, 58, 80, 82
   concentrated   58, 80, 82
   production, uses   80, 82
superoxide   15

termination   40
test
   for acid functional group
      (O-H)   57, 58
   for alcohol (O-H)   57, 58
   for alkenes   45
   for anions   87
   for cations   87
   for chloride ions   21
   for halide ions   21
   for haloalkane   90
   for -OH group   57, 58
   for sulfate ions   87
tetrafluoroethene   46
thermal cracking   45
thermal decomposition   16, 86
titanium   21, 81
titration
   calculations   99
trans   34
transition
   metal   81
   states   68, 70
trans-platin   95

ultraviolet
   light   33, 40, 44
unsaturated   44

vacuum filtration   87
valence shell
   electron   8
   electron pair repulsion
      (VSEPR) theory   8
Van der Waals forces   4, 20, 50
vanadium   82
vegetable oil   45
vinegar   38, 57
vinyl chloride   46
volume   93, 94, 98, 99
VSEPR theory   8

wavenumber   58
wine   38, 57

yeast   56, 81
yield   75, 76, 82, 94

zinc   22

Periodic Table

Key

Molar mass g mol⁻¹
Symbol
Name
Atomic number

Period	Group 1	Group 2												Group 3	Group 4	Group 5	Group 6	Group 7	Group 8
1	1 **H** Hydrogen 1																		4 **He** Helium 2
2	7 **Li** Lithium 3	9 **Be** Beryllium 4												11 **B** Boron 5	12 **C** Carbon 6	14 **N** Nitrogen 7	16 **O** Oxygen 8	19 **F** Fluorine 9	20 **Ne** Neon 10
3	23 **Na** Sodium 11	24 **Mg** Magnesium 12												27 **Al** Aluminium 13	28 **Si** Silicon 14	31 **P** Phosphorus 15	32 **S** Sulphur 16	35.5 **Cl** Chlorine 17	40 **Ar** Argon 18
4	39 **K** Potassium 19	40 **Ca** Calcium 20	45 **Sc** Scandium 21	48 **Ti** Titanium 22	51 **V** Vanadium 23	52 **Cr** Chromium 24	55 **Mn** Manganese 25	56 **Fe** Iron 26	59 **Co** Cobalt 27	59 **Ni** Nickel 28	63.5 **Cu** Copper 29	65.4 **Zn** Zinc 30		70 **Ga** Gallium 31	73 **Ge** Germanium 32	75 **As** Arsenic 33	80 **Se** Selenium 34	80 **Br** Bromine 35	84 **Kr** Krypton 36
5	85 **Rb** Rubidium 37	88 **Sr** Strontium 38	89 **Y** Yttrium 39	91 **Zr** Zirconium 40	93 **Nb** Niobium 41	96 **Mo** Molybdenum 42	(99) **Tc** Technetium 43	101 **Ru** Ruthenium 44	103 **Rh** Rhodium 45	106 **Pd** Palladium 46	108 **Ag** Silver 47	112 **Cd** Cadmium 48		115 **In** Indium 49	119 **Sn** Tin 50	122 **Sb** Antimony 51	128 **Te** Tellurium 52	127 **I** Iodine 53	131 **Xe** Xenon 54
6	133 **Cs** Caesium 55	137 **Ba** Barium 56	139 **La** Lanthanum 57	178 **Hf** Hafnium 72	181 **Ta** Tantalum 73	184 **W** Tungsten 74	186 **Re** Rhenium 75	190 **Os** Osmium 76	192 **Ir** Iridium 77	195 **Pt** Platinum 78	197 **Au** Gold 79	201 **Hg** Mercury 80		204 **Tl** Thallium 81	207 **Pb** Lead 82	209 **Bi** Bismuth 83	210 **Po** Polonium 84	210 **At** Astatine 85	222 **Rn** Radon 86
7	223 **Fr** Francium 87	226 **Ra** Radium 88	227 **Ac** Actinium 89																

140 **Ce** Cerium 58	141 **Pr** Praseodymium 59	144 **Nd** Neodymium 60	(147) **Pm** Promethium 61	150 **Sm** Samarium 62	152 **Eu** Europium 63	157 **Gd** Gadolinium 64	159 **Tb** Terbium 65	163 **Dy** Dysprosium 66	165 **Ho** Holmium 67	167 **Er** Erbium 68	169 **Tm** Thulium 69	173 **Yb** Ytterbium 70	175 **Lu** Lutetium 71
232 **Th** Thorium 90	(231) **Pa** Protactinium 91	238 **U** Uranium 92	(237) **Np** Neptunium 93	(242) **Pu** Plutonium 94	(243) **Am** Americium 95	(247) **Cm** Curium 96	(245) **Bk** Berkelium 97	(251) **Cf** Californium 98	(254) **Es** Einsteinium 99	(253) **Fm** Fermium 100	(256) **Md** Mendelevium 101	(254) **No** Nobelium 102	(257) **Lr** Lawrencium 103

www.ingramcontent.com/pod-product-compliance
Lightning Source LLC
Chambersburg PA
CBHW081418300426
44109CB00019BA/2341